CARBON FIBERS

INTERNATIONAL FIBER SCIENCE AND TECHNOLOGY SERIES

Volume 1: Handbook of Fiber Science and Technology: Chemical Processing of Fibers and Fabrics – Fundamentals and Preparation (in two parts), *edited by Menachem Lewin and Stephen B. Sello*

Volume 2: Handbook of Fiber Science and Technology: Chemical Processing of Fibers and Fabrics – Functional Finishes (in two parts), *edited by Menachem Lewin and Stephen B. Sello*

Volume 3: Carbon Fibers, *by Jean-Baptiste Donnet and Roop Chand Bansal*

Other Volumes in Preparation

CARBON FIBERS

Jean-Baptiste Donnet
Centre National de la Recherche Scientifique
Physico-Chimie des Surfaces Solides
Mulhouse, France

Roop Chand Bansal
Panjab University
Chandigarh, India

MARCEL DEKKER, INC. New York and Basel

Library of Congress Cataloging in Publication Data

Donnet, Jean-Baptiste, 1923–
Carbon fibers.

(International fiber science and technology series ; 3)
Includes bibliographical references and indexes.
1. Carbon fibers. I. Bansal, Roop Chand, 1937- II. Title. III. Series.
TA419.8.F5D66 1984 620.1'97 83-26268
ISBN 0-8247-7098-6

MARCEL DEKKER, INC.
270 Madison Avenue, New York, New York 10016

Current printing (last digit):
10 9 8 7 6 5 4 3 2 1

PRINTED IN THE UNITED STATES OF AMERICA

About the Series

When human life began on this earth *food* and *shelter* were the two most important necessities. Immediately thereafter, however, came *clothing.* The first materials used for it were fur, hide, skin, and leaves—all of them sheetlike, two-dimensional structures not too abundantly available and somewhat awkward to handle. It was then—quite a few thousand years ago—that a very important invention was made: to *manufacture* two-dimensional systems—fabrics—from simple monodimensional elements—fibers; it was the birth of textile industry based on fiber science and technology. Fibers were readily available everywhere; they came from animals (wool, hair, and silk) or from plants (cotton, flax, hemp, and reeds). Even though their chemical composition and mechanical properties were very different, yarns were made of the fibers by spinning and fabrics were produced from the yarns by weaving and knitting. An elaborate, widespread, and highly sophisticated art developed in the course of many centuries at locations all over the globe virtually independent from each other. The fibers had to be gained from their natural sources, purified and extracted, drawn out into yarns of uniform diameter and texture, and converted into textile goods of many kinds. It was all done by hand using rather simple and self-made equipment and it was all based on empirical craftsmanship using only the most necessary quantitative measurements. It was also performed with no knowledge of the chemical composition, let alone the molecular structure of the individual fibers. Yet by ingenuity, taste, and patience, myriads of products of breathtaking beauty, remarkable utility, and surprising durability were obtained in many cases. *This first era* started at the very beginning of civilization and extended into the twentieth century when steam-driven machinery invaded the mechanical operations and some empirical procedures—mercerization of cotton, moth-proofing of wool and loading of silk—started to introduce some chemistry into the processing.

The second phase in the utilization of materials for the preparation and production of fibers and textiles was ushered in by an accidental discovery which Christian Friedrich Schoenbein, chemistry professor at the University of Basel in Switzerland, made in 1846. He observed that cotton may be converted into a soluble and plastic substance by the action of a mixture of nitric and sulfuric acid; this substance or its solution was extruded into fine filaments by Hilaire de Chardonnet in 1884.

Organic chemistry, which was a highly developed scientific discipline by that time, gave the correct interpretation of this phenomenon: the action of the acids on cellulose—a natural fiber former—converted it into a *derivative,* in this case into a cellulose nitrate, which was soluble and, therefore, spinnable. The intriguing possibility of manipulating natural products (cellulose, proteins, chitin, and others) by chemical action and thereby rendering them soluble, resulted in additional efforts which led to the discovery and preparation of several cellulose esters, notably the cellulose xanthate and cellulose acetate. Early in the twentieth century each compound became the basis of a large industry: viscose rayon and acetate rayon. In each case special processes had to be designed for the conversion of these two compounds into a fiber, but once this was done, the entire mechanical technology of yarn and fabric production which had been developed for the natural fibers was available for the use of the new ones. In this manner new textile goods of remarkable quality were produced, ranging from very shear and beautiful dresses to tough and durable tire cords and transport belts. Fundamentally these materials were not truly "synthetic" because a known natural fiber former—cellulose or protein—was used as a base; the new products were "artificial" or "man-made." In the 1920s, when viscose and acetate rayon became important commercial items polymer science had started to emerge from its infancy and now provided the chance to make *new fiber formers* directly by the polymerization of the respective monomers. Fibers made out of these polymers would therefore be "truly synthetic" and represent additional, extremely numerous ways to arrive at new textile goods. Now started the *third era* of fiber science and technology. First the basic characteristics of a good synthetic fiber former had to be established. They were: ready spinnability from melt or solution; resistence against standard organic solvents, acids, and bases; high softening range (preferably above 220°C); and the capacity to be drawn into molecularly oriented fine filaments of high strength and great resilience. There exist literally many hundreds of polymers or copolymers which, to a certain extent, fulfill the above requirements. The first commercially successful class were the *polyamides*, simultaneously developed in the United States by W. H. Carothers of duPont and by Paul Schlack of I. G. Farben in Germany. The *nylons*, as they are called commercially, are still a very important class of textile fibers covering a remarkably wide range of properties and uses. They were soon (in the 1940s) followed by the *polyesters*, *polyacrylics*, and *polyvinyls*, and somewhat later (in the 1950s) there were added the *polyolefins* and *polyurethanes*. Naturally, the existence of so many fiber formers of different chemical composition initiated successful research on the molecular and supermolecular structure of these systems and on the dependence of the ultimate technical properties on such structures.

As time went on (in the 1960s), a large body of sound knowledge on structure-property relationships was accumulated. It permitted embarkation on the reverse approach: "tell me what properties you want and I shall *tailor-make* you the fiber former." Many different techniques exist for the "tailor-making": graft and block copolymers, surface treatments, polyblends, two-component fiber spinning, and cross-section modification. The systematic use of this "macromolecular engineering" has led to a very large number of *specialty fibers* in each of the main classes; in some cases they have properties which none of the prior materials—natural and "man-made"—had, such as high elasticity, heat setting, and moisture repellency. An important result was that the new fibers were not content to fit into the existing textile machinery, but they suggested and introduced substantial modifications and innovations such as modern high-speed spinning, weaving and knitting, and several new technologies of texturing and crimping fibers and yarns.

This third phase of fiber science and engineering is presently far from being complete, but already a *fourth era* has begun to make its appearance, namely in fibers for uses *outside* the domain of the classical textile industry. Such new applications involve fibers for the reinforcement of thermoplastics and duroplastics to be used in the construction of spacecraft, airplanes, buses, trucks, cars, boats, and buildings; optical fibers for light telephony; and fibrous materials for a large array of applications in medicine and hygiene. This phase is still in its infancy but offers many opportunities to create entirely new polymeris systems adapted by their structure to the novel applications outside the textile fields.

This series on fiber science and technology intends to present, review, and summarize the present state in this vast area of human activities and give a balanced picture of it. The emphasis will have to be properly distributed on synthesis, characterization, structure, properties, and applications.

It is hoped that this series will serve the scientific and technical community by presenting a new source of organized information, by focusing attention to the various aspects of the fascinating field of fiber science and technology, and by facilitating interaction and mutual fertilization between this field and other disciplines, thus paving the way to new creative developments.

Herman F. Mark

Preface

Carbon fibers are a new breed of high strength material. Although initially developed for the aerospace industry and for military aircraft, they are now finding wide application in commercial and civilian aircraft, recreational, industrial, and transportation markets. They have low density and thus possess the highest specific strength and stiffness of all currently available engineering materials. They are used in composites with a light weight matrix, generally an epoxy resin, occasionally a polyester of polyimide, and now frequently carbon. Their use as reinforcing agents in light metals such as aluminum and magnesium is expected to be a future development. Carbon fiber composites are ideally suited to applications where strength, stiffness, lower weight, and outstanding fatigue characteristics are critical requirements. They are also finding applications where high temperature, chemical inertness, and high damping are important. Thus a tremendous amount of research activity has been directed in the carbon fiber field.

The Plastics Institute, London, arranged two international conferences specifically devoted to investigations in the carbon fiber field. The Royal Society of London and the American Chemical Society initiated discussion on high-performance fibers, including carbon fibers. In addition, International Conferences on Carbon, held biennially in the United States and occasionally by the German Carbon group, and the International Conferences on Carbon and Graphite held under the auspices of the Society of Chemical Industry, London, the United States, and European SAMPE meetings have devoted several special sessions to contributions related to carbon fibers. Research papers concerning carbon fibers appear in such diverse international journals as *Journal of Materials Science*, *Journal of Polymer Science*, *Journal of Colloid and Interface Science*, *Carbon*, *Fuel*, *Tanso*, and many more. The literature on carbon fibers is, therefore, widely scattered.

The object of this book is to bring together in one volume the very significant contributions to the carbon fiber field and to present the diverse opinions and views expressed by various investigators concerning the chemistry of preparation (Chapter 1), structure (Chapter 2), surface properties (Chapter 3), mechanical and other physical properties (Chapter 4), and applications (Chapter 5) of carbon fibers. The book, however, is not intended to be a complete review of the subject. A considerable amount of literature on carbon fibers is classified and is thus not accessible. Attempts have been made to collect literature from all available sources and to refer to the latest contributions. It is hoped that this book will serve as a useful guide to our colleagues in the carbon field in industry, research institutes, and universities.

The authors express their thanks and gratitude to all who helped, particularly Professor E. Fitzer for his valuable comments and suggestions. We also express our thanks to our colleagues of the laboratory, Dr. P. Ehrburger for the benefit of discussion during preparation of the manuscript, Dr. G. Majowski for documentation, reading the proofs, and for translation of papers from Russian, Japanese, and German; to F. Müller for photography, C. Dehr for drawings, and Mrs. G. Wachtel, S. Wagner, and F. Sorgato for typing. Dr. R. C. Bansal is also thankful to the DGRST and CNRS France for their financial support for his stay, and the Panjab University, Chandigarh, India for granting sabbatical leave. To Suzanne and Rajesh, our wives, we are grateful for their love and spiritual challenge to us.

Jean-Baptiste Donnet
Roop Chand Bansal

Contents

CARBON FIBERS

1

Preparation of Carbon Fibers

Carbon fibers had been known for nearly a century when Thomas Edison first used them as filaments for incandescent lamps. These early carbon fibers were prepared by the carbonization of bamboo and rayon. Soon after the development of tungsten filaments this use of carbon fibers was abandoned. Later carbon fibers were prepared from rayon alone, to be used for thermal insulation, but, these fibers could not compete with glass fibers.

The present upsurge in the development of carbon fibers stems from work carried out in the 1950s at Wright-Patterson Air Force Base where the first major effort was undertaken to make high-strength carbon fibers. Union Carbide [1] developed this process and was able to enhance the mechanical properties of rayon-based carbon fibers using stress graphitization. This process led to the production of high-modulus and high-strength carbon fibers and to Union Carbide's introduction of Thornel 25, 40, and 50 carbon fibers.

At about the same time, Shindo in Japan [2] and Watt and co-workers in the United Kingdom [3], working independently, successfully prepared carbon fibers from polyacrylonitrile (PAN). These carbon fibers had relatively high modulus but low strength. The mechanical properties of these PAN-based carbon fibers were further enhanced by carrying out certain oxidative treatments under stress. The PAN process of carbon fiber production received the immediate attention of several manufacturers because the yield of the carbon fiber was almost double that of its yield from rayon. Moreover, the process did not entail the difficulties and high cost of high-temperature stressing treatment.

Kureha Chemical Industry Company in Japan introduced low-to intermediate-strength and intermediate-modulus carbon fibers prepared from certain specific fractions of asphalt or pitch materials [4] and lignin and lignosulfonates [5]. These carbon fibers could be prepared at low cost.

These developments gave incentive to investigators in the carbon processing field to search for newer and low cost sources for carbon fiber production. The need to produce low cost carbon fibers has recently been given further impetus with the realization that carbon fibers may be used successfully to produce lightweight automobile parts that reduce energy consumption. Thus, several types of precursor materials have been investigated and their feasibility for large-scale production of carbon fibers assessed.

The first pitch-based carbon fiber was developed by Otani [15,16] in Japan who used polyvinyl chloride (PVC) pitch as the raw material. Petroleum pitch is the raw material currently used for producing pitch-based carbon fibers. Kureha Chemical Company of Japan has built a commercial plant with a production capacity of 250 tons per year of carbon fibers from petroleum pitch.

Union Carbide has been working to develop a process [6,7] to manufacture carbon fibers using certain types of mesophase pitches, and has recently completed a new mesophase pitch-based carbon fiber plant at Greenville, South Carolina. This plant attained its maximum capacity of about 230 tons per year in March 1981, and is producing a variety of continuous yarn of 2000 and 4000 filaments with elastic modulus varying between 170 and 830 GPa [7a]. Tables 1-3 record the mechanical properties of various types of carbon fibers which are being produced commercially at the present time.

During the past 20 years several other high-performance materials, such as boron and Kevlar, have become available as continuous fibers. These are all of low density and are of interest owing to their favorable specific properties. The specific properties of carbon fibers are compared with other high-performance materials developed recently and with other more conventional materials in Table 4.

1.1 PRECURSORS FOR CARBON FIBERS

1.1.1 Polyacrylonitrile (PAN) as Precursors

Several types of acrylic polymers with small variations in their composition have been used for the production of carbon fibers. The exact composition of a particular precursor is part of the patent literature. However, the general composition contains a minimum of 85% acrylonitrile and a maximum of 15% comonomer such as methyl methacrylate, methyl acrylate, vinyl acetate, vinyl chloride, and other monovinyl compounds [8]. The influence of precursor composition will be discussed a little later in the chapter. A British patent recommends the use of Courtelle which is suggested to be a terpolymer of acrylonitrile, methyl acrylate, and another vinyl monomer. A limitation in the conversion of PAN into carbon fibers arises from the slow oxidation which is essential to stabilize the oriented structure and to stabilize the fiber against fusion during the high temperature treatment. Thus, a PAN precursor fiber which has fine filaments is considered a more

Table 1 Polyacrylonitrile-Based Carbon Fibers Available Commercially

Name of fiber	Manufacturer	Modulus (GN m^2)	Strength (GN m^2)
High modulus			
Celion Gy-70	Celanese (U.S.)	517	2.14 (1.86)
Celion Gy-50	Celanese (U.S.)	345	—
Thermolon -T3	Tokai Electrode Japan	483	1.68
Torayaca M-40A	Toraya (Japan)	345 (422)	2.45
Grafil HM	Courtaulds (U.S.)	360	2.12 (2.34)
Magnamite HMS	Hercules (U.S.)	407	2.17
Rigilor AGT	France	422	2.18
Fortafil CGS	Great Lakes (U.S.)	310 (331)	2.41 (2.76)
Modmer I (discontinued)	Morganite Modmer (U.S., U.K.)	414	2.06
Sigrifil HM	Sigri (West Germany)	343 (483)	1.96 (1.48)
Rigilor 250	Le Carbone Lorraine (France)	379	1.72
Rigilor 350	Le Carbone Lorraine (France)	310	2.41
Polycarbon M	Polycarbon U.S.	414	2.06
High strength			
Carbolon Z2	Japan	276	3.02
Grafil HTS	Courtaulds (U.K.)	250	2.60
Modmer II (discontinued)	Morganite Modmer (U.K.)	230	2.42
Sigrifil HF	Sigri (West Germany)	216	2.35
Panex 30A	Stackpole (U.S.)	245	3.04
Panex-20	Stackpole (U.S.)	145	3.24
HTA-3000	Stackpole (Japan)	215	2.63
Torayaca T300	Toraya (Japan)	245	3.04
Thornel-300	Union Carbide (U.S.)	234	2.51
Celion 1000	Celanese (U.S.)	234	2.48
Celion 3000	Celanese (U.S.)	234	2.76
Celion 6000	Celanese (U.S.)	234	2.76
Grafil A	Courtaulds (U.K.)	214	2.69
Grafil HT	Courtaulds (U.K.)	248	2.76
Fortafil 3U	Great Lakes (U.S.)	186	2.48
Fortafil 3	Great Lakes (U.S.)	207	2.48
Magnamite AS	Hercules (U.S.)	221	3.10

Table 1 (Continued)

Name of fiber	Manufacturer	Modulus (GN m^2)	Strength (GN m^2)
Magnamite HTS	Hercules (U.S.)	248	2.76
Polycarbon T	Polycarbon (U.S.)	276	2.41
Thornel -T400	Union Carbide (U.S.)	207	3.10
Thornel T 300A	Union Carbide (U.S.)	241	2.48
Other types			
Carbolon -A	Japan	99	1.19
Carbolon -Ag	Japan	119	0.99
Hyfil 27	U.K.	290	2.21
Grafil As	Courtaulds (U.K.)	205	2.30
Rigilor ACT	France	226	2.26
Modmer III (discontinued)	Morganite Modmer (U.K.)	200 (234)	2.34 (2.41)
Sigrifil NF	Sigri (West Germany)	177	1.96
Panex 30 C	Stackpole (U.S.)	227	2.96 (2.31)
Torayaca -T200	Toray (Japan)	245	2.25
Polycarbon A	Polycarbon (U.S.)	241	2.06
Paney 30 y	Stackpole (U.S.)	221	1.79

Source: Data from Ref. 8 and Ref. 161.

suitable fiber because it requires less time for oxidative stabilization. Ezekiel [9] has shown that low tenacity homopolymers can be successfully carbonized and graphitized into high-modulus graphite fibers. However, it has been felt that increased tension exposure during graphitization generally yields carbon fibers with good mechanical properties using any PAN precursor material [8]. However, this approach has significant process implications and is generally avoided. A brief review of the various PAN precursor materials for the production of carbon fibers has been published by Ross [10].

Although PAN fibers are more expensive than rayon fibers, they are extensively used as a source of carbon fibers because their carbon yield is almost double that of the carbon yield from rayon. The commercial potential of PAN-based fibers can probably best be judged by the number of manufacturers who are actively competing in the market (Table 1).

Table 2 Rayon-Based Carbon Fibers Available Commercially

Carbon fiber	Manufacturer	Modulus (GN m^2)	Strength (GN m^2)
High modulus			
Thornel 50	Union Carbide (U.S.)	345	2.07
Thornel 75	Union Carbide (U.S.)	517	2.52
HMG-50	Hitco (U.S.)	345	2.07
High strength			
Thornel 40	Union Carbide (U.S.)	276	1.72
HMG-40	Hitco (U.S.)	276	1.72
Other types			
Thornel	Union Carbide (U.S.)	172	1.24
VyB-70 1/2	Union Carbide (U.S.)	41	0.83
WyB-85 1/2	Union Carbide (U.S.)	34	0.62
HMG-25	Hitco (U.S.)	172	1.03
Gy2-1	Hitco (U.S.)	28	1.38
Cx5/Cx2	Polycarbon (U.S.)	28	1.21
GSCy-2-5	Carborundum (U.S.)	28	0.97

Source: Data from Ref. 8 and Ref. 161.

1.1.2 Cellulosic Precursors

Cellulosic precursor fibers, although of considerable historic significance, are no longer an important source of carbon fibers. Hundreds of patents have been issued and considerable published data exist on these precursor materials. These precursor fibers include cotton, linen, ramie, sisal, heme, and flax, although regenerated cellulose or rayon fibers have been more commonly used. Polynosic fibers have been suggested, but they do not provide any real advantage. More recent investigations into the production of carbon fibers from cellulosic fibers are concentrated toward modifying the degradation mechanism so that the carbon yield can be enhanced. A degree of success has already been achieved in this direction by carrying out pyrolysis

Table 3 Pitch-Based and Other Carbon Fibers Available Commercially

Carbon fiber	Manufacturer	Modulus (GN m2)	Strength (GN m2)
Type P pitch-based	Union Carbide (U.S.)		
High modulus		380–830	1.72–2.41
High strength		193–241	2.07–2.76
Kureha pitch-based	Kureha (Japan)		
High modulus		262	0.52
High strength		28	1.72
Kaya carbon			
Lignin based		–	0.59

Source: Data from Ref. 8 and Ref. 161.

of cellulosic precursors in the presence of certain inorganic compounds, or by modification of cellulose, or by varying the reactive environment [10–13]. In addition, stress graphitization has been resorted to by the Union Carbide investigators [14] in an effort to obtain a significant increase in the mechanical properties of these carbon fibers.

1.1.3 Pitch Precursors

Certain fractions of asphalt and pitch materials were converted into carbon and graphite fibers by heat treatment, extrusion, spinning followed by stabilization, carbonization, and graphitization processes [4]. The carbon fibers thus obtained had intermediate strength and modulus properties. This led Union Carbide investigators to develop a process for the preparation of high-modulus carbon fibers from polynuclear aromatic pitches [7]. These pitch carbon fibers, however, had poor strength properties. Another shortcoming of these fibers was the nonuniformity and the nonreproducibility of their properties. At the present time Union Carbide is working on a process for the large-scale production of carbon fibers from mesophase pitches [6,7]. These carbon fibers have been found to have mechanical properties similar to those obtained from PAN precursors. Otani [15,16] investigated the possibility of converting PVC-derived pitches into carbon fibers and improving their properties by using tension during their heat treatment. However, this preparation procedure involves high temperature stretching which is a difficult and expensive treat-

Table 4 Comparison of Mechanical Properties of Carbon Fibers with Other Fibrous Reinforcing Materials

Fiber	Specific gravity (SG)	Ultimate tensile strength (UTS) (GN m2)	Relative specific strength (UTS/SG)	Young's modulus (E) (GN m2)	Relative Young's modulus (E/SG)
Carbon fiber Type I	1.90	2.6	1.37	340	179
Carbon fiber Type II	1.77	2.9	1.64	240	136
Carbon Fiber Type III	1.76	2.6	1.48	190	108
E-glass	2.59	3.44	1.38	72	29
S-glass	2.49	2.60	1.06	80	32
Kevlar 29	1.44	2.7	1.80	60	42
Kevlar 49	1.45	2.7	1.86	130	90
Boron (on tungsten)	2.50	3.75	1.50	400	160
Asbestos crocidolite	3.40	3.50	1.03	190	58
Asbestos chrysotite	2.50	3.11	1.24	160	64
Silicon carbide (on tungsten)	3.50	3.50	1.00	400	114

Source: Data from Refs. 162 and 163.

ment as compared to the Union Carbide process. Pitch carbon fibers have a future commercial potential owing to the low cost of the source material and better carbon yield. It may, however, be noted that in the case of pitch, although the mass yield in graphitization is substantially higher than for PAN, the overall yield from raw pitch is lower due to losses in the mesophase conversion step.

1.1.4 Nonheterocyclic Aromatic Polymer Precursors

The possibility of converting nonheterocyclic aromatic polymers such as phenolic polymers [17–19], phenol formaldehyde resin [20–22], polyacenaphthalene [23,24], polyacrylether [25], certain polyamides [26–29], and polyphenylene [30], into carbon fibers has also been investigated. These substances have several advantages, such as easy cyclization into a graphitic structure, easy elimination of noncarbon atoms, and a greater carbon yield.

Economy and Wohrer [17] and Lin and Economy [18] used "kynal," a phenolic textile fiber produced by Carborundum, whereas Kawamura and Jenkins [19] used phenolic hexamine polymers for the preparation of glassy carbon fibers. The carbon yield varied between 60-70%. Carbon fibers with poor mechanical properties were also obtained by the carbonization of phenolformaldehyde [22-24]. These carbon fibers can be used as adsorbents after activation [30].

Polyacenaphthalene and polyindene on dehydrogenation yield a fully aromatic polymer and, therefore, have been suggested [23] as potential precursors for carbon and graphite fibers. Pitch obtained from acenaphthalene has also been used for preparing carbon fibers [24]. Controlled oxidation followed by carbonization of polyacrylether [25] results in the formation of carbon fibers with a good yield. Certain polyamides prepared by the condensation of acrylene diamines and dicarboxylic acids were also converted into carbon fibers after oxidative stabilization [26–29].

General Electric Co. has recently patented a continuous method for making polyacetylene fibers which have been converted into carbon fibers [31–34]. The polyacetylene fiber is heat-stabilized below 1000°C in air or inert atmosphere and then carbonized and graphitized in an inert atmosphere, producing carbon fibers with tensile strength of 1600–2170 MPa and a modulus range of 350–490 GPa [8].

1.1.5 Aromatic Heterocyclic Polymers as Precursors

Certain high temperature-resistant organic polymers with aromatic heterocyclic structures have also been considered as precursors for carbon fibers. These polymers are generally linear and have a relatively high aromatic content. Their degradation mechanism is similar to that of PAN, producing a ladder polymer-type structure. They generally produce anisotropic carbon fibers in good yield. The only drawback with these polymers is their high cost, although they can produce carbon fibers with unique mechanical properties without any modification [8]. Among this class of polymers, polyimides [35,36], polybenzimidazole [35,27], polybenzimidazonium salt [37], and polytriadiazoles [38], have been successfully converted into high-modulus carbon fibers. In some cases the carbon yield has been found to be as high as 90% [37].

1.1.6 Linear Polymers as Precursors

Certain linear thermoplastic polymers have been investigated for the preparation of carbon fibers. For example, mixtures of polyethylene and polypropylene and several aromatic distillation residues were melt-spun after heat treatment at 420°C, oxidized, and then carbonized to obtain carbon fibers [39]. Polyvinyl chloride, another member of this class of polymers, has been converted into carbon fibers, either by direct pyrolysis under controlled conditions [40], or by first converting into PVC pitch which was melt-spun and subsequently carbonized and graphitized [15,16]. Copolymers of vinyl chloride and vinyl alcohol were treated with concentrated sulfuric acid to render the fiber infusible and then heated to produce carbon fibers [4]. A process for the cyclization and graphitization of fibers has also been described.

Tomizuka et al. [156] have recently obtained carbon fibers with 40% yield by heating high modulus polyamide (Kevlar) fibers in an inert atmosphere. The carbon fibers heat-treated to 1000°C were circular in cross section and isotropic over the entire area, while those prepared above 2000°C were both isotropic circular and anisotropic irregular in shape and cross-section. The carbon fibers are nongraphitizable and contain a large number of micro and transitional voids. They have a strength equivalent and modulus double that of the commercial low-modulus carbon fibers.

Carbon fibers have also been obtained from polymethyl vinyl ketone fibers [42]. These fibers are oriented by stretching, cyclized by dehydration, and carbonized in inert atmosphere. The mechanism of conversion into carbon fibers is similar to that of PAN, but the mechanical properties are poor. Polyvinyl alcohol and polyvinyl acetate were also converted into carbon fibers [43] with low tensile strength. The carbon fiber yield is also poor, but has been found to be enhanced by the addition of ammonium salts of phosphoric acid [44].

1.1.7 Coal as a Precursor

Carbon fibers have also been reported to be prepared from certain coal extracts [45,46] obtained by digesting coal at elevated temperatures and under high pressures with a high-boiling aromatic solvent. The coal extract of a suitable consistency was spun into fibers by the melt-spinning technique. The coal fibers were rendered infusible by oxidation in air at 250°C and then carbonized up to 1000°C in an inert atmosphere. The fibers were found to have a circular cross section and uniform diameter with a smooth external surface. They were, however, reported to be isotropic and very poor in strength [46].

It appears that the significant factors in obtaining a high carbon yield are: the nature of the polymer precursor, the nature of the degradation process, the capacity of the precursor for cyclization, ring fusion and coalescence, and the nature of the stabilizing pretreatment. Table 5 shows the result of thermogravimetric analysis

Table 5 Weight Loss on Heating at 1000°C in Helium for Different Fibers

Polymer type	Weight loss total (%)
Rayons	87–89
Polyacrylonitrile	60,67
Polyacrylonitriles, preoxidized	38
PRD-14	52
X-101	53
Saran	74
Poly(vinyl alcohol)	93
Ramie	91
Pitch	≅ 30

Source: Data from Ref. 35.

(TGA), carried out on several polymeric fibers at 1000°C in helium atmosphere, and indicates that polymerization of PAN precursor yields the maximum amount of carbon fiber. The carbon yield in the case of pitch precursors is still higher.

Another condition for the suitability of a precursor material for carbon fiber formation is that the degradation of the precursor should involve a cyclization of a mesophase type of mechanism. Similarly, the glass transition temperature of the precursor, or its stabilized intermediate form, is a critical parameter during carbonization and graphitization processes.

Thus, three precursors which are being used for the large-scale production of carbon fibers are PAN, rayon, and pitches. The details of the processes used for different precursors are not the same, although their essential features are similar. They involve a stabilizing treatment to prevent melting or fusion of the fiber, a carbonizing heat treatment to eliminate the noncarbon elements, and a high-temperature graphitizing treatment to enhance the mechanical properties of the final carbon fiber. The term "graphitizing" is actually a misnomer because care is always taken to avoid the formation of a truly graphitic structure. In fact, the materials used for the preparation of carbon fibers are generally nongraphitizing. Possibly, the term "graphitic fibers" has been used perhaps to describe the turbostratic structure of the high-modulus carbon fibers.

In the case of PAN precursors, the low-temperature stabilization treatment is generally given in air for several hours at about 220°C. Two different approaches have been described by Shindo [2] and by Johnson and co-workers [48] and Moreton et al. [47], the essential difference being the prevention of shrinkage by stretching during this low-temperature treatment. The black fibers thus obtained are

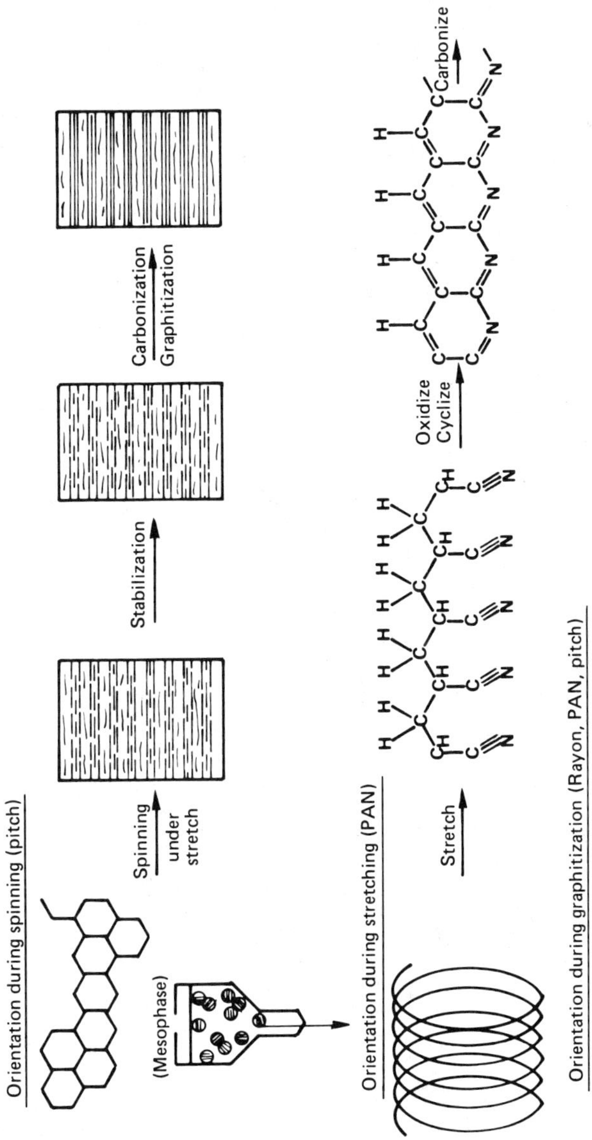

Figure 1 Methods for creating preferred orientation in carbon fibers. (From Ref. 160, photo courtesy of E. Fitzer.)

then carbonized in an inert atmosphere at 1000-1500°C with a slow heating rate in order to avoid damage to the high degree of molecular order present in the fiber. The fibers are then given a graphitizing treatment at temperatures above 2000°C and up to 3000°C to improve the texture of the fiber and to enhance the Young's modulus. The strength and the modulus of these fibers can also be improved by hot stretching.

Rayon fibers also involve a low-temperature stabilizing treatment in air and carbonization at temperatures higher than 1000°C. However, the graphitization which is usually carried out above 2000°C must be performed while the fibers are stretched. The production of rayon fibers is thus expensive because of its low yield and the costly high-temperature stretching treatment.

The pitch fibers also involve the same sequence of thermal stabilization by oxidation, carbonization, and graphitization. The resulting fiber, however, has poor mechanical properties, which can be improved by graphitization under stress. In order to avoid the expensive high-temperature stress treatment, mesophase pitch fibers have been used. The orientation of the fibers is carried out by spinning the pitch under stress at low temperature. These fibers retain their orientation during the subsequent stabilization and carbonization treatments to produce highly oriented high-modulus carbon fibers without any stress graphitization.

Thus, in order to produce carbon fibers with a high degree of preferred orientation along the fiber axis, the precursor fibers are given a stretching treatment during one of the stages of the pyrolysis process (Fig. 1).

Glassy carbon fibers with relatively low modulus can be produced without any preoxidation treatment from certain phenolic resins [19, 49,50]. These fibers may be used profitably where chemical inertness and low price are of greater significance than very good mechanical properties. We will now discuss in greater detail the chemistry of various processes used commercially for the preparation of carbon fibers.

1.2 PREPARATION OF CARBON FIBERS FROM POLYACRYLONITRILE

PAN fiber is recognized as the most important and promising precursor for the present manufacture of carbon fibers. The advantages are a high degree of molecular orientation, higher melting point, and a greater yield of the carbon fiber. PAN fibers form a thermally stable, highly oriented molecular structure when subjected to a low-temperature heat treatment, which is not significantly disrupted during the carbonization treatment at higher temperatures, giving rise to a carbon fiber with good mechanical properties. An overall process schematic for the preparation of carbon fiber from PAN is shown in Fig. 6.

The acrylonitrile molecule has a nitrile group which is highly polar

acrylonitrile

(I)

so that polymerization can take place in the presence of free radicals or negative initiators, leading to the production of polyacrylonitrile

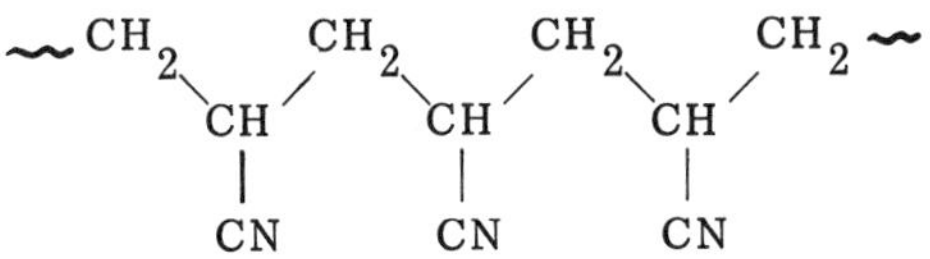

polyacrylonitrile

(II)

The highly polar character of the nitrile group causes strong dipole-dipole forces which act as crosslinks, making the polymers soluble only in highly ionizing solvents, increasing its melting point, and making it more suitable as a carbon fiber precursor. However, these dipole-dipole interactions prevent necessary rearrangements for the development of crystallinity and cause shrinkage [51–55] of the fiber on heat treatment. This shrinkage of the PAN fiber is disadvantageous to the carbon fiber manufacturer because it disturbs the orientation of the fiber, thus producing carbon fiber with poor mechanical properties. Steps must therefore be taken to prevent this shrinkage. Shindo [2], who did not prevent shrinkage, obtained carbon fibers with poor mechanical properties, whereas Johnson et al. [48] and Watt and Johnson [56], who prevented shrinkage by stretching the fiber on a frame, produced carbon fibers with better mechanical properties.

The basic principles involved in the conversion of PAN fibers into carbon fibers are a treatment of the fibers in an oxidizing atmosphere at temperatures in the range 200-300°C while they are under tension in order to prevent shrinkage or even cause extension. This treatment converts thermoplastic PAN into a nonplastic cyclic or ladder compound capable of withstanding the high temperature present in the next part of the process. The oxidized fibers are then carbonized to about 1000°C without tension and further heated to any temperature up to 3000°C depending upon the required Young's modulus of the final carbon fiber (Table 6).

Table 6 Tensile Strength and Young's Modulus of Carbon Fibers as a Function of Heat-Treatment Temperature

	Ultimate tensile strength		Young's modulus	
Temp. of treatment (°C)	Shindo ($GN\ m^2$)	Watt and Johnson ($GN\ m^2$)	Shindo ($GN\ m^2$)	Watt and Johnson ($GN\ m^2$)
1000	0.75	1.4	112	161
1500	0.68	2.2	119	238
2000	0.50	1.8	168	300
2500	0.46	1.8	133	384
2800	–	1.5	–	426
3000	0.24	–	147	–

Source: Data from Ref. 56.

1.2.1 Thermal Stabilization of PAN Fibers

The unsaturated side groups present in the linear molecule of polyacrylonitrile can cause additional polymerization reactions, which depend upon the physical state of the polymer, the presence of additives such as comonomers, byproducts and traces of initiators from the original polymerization, end groups, and ionic groups [57–59], which may have been introduced during spinning of the fiber and the rate of heating. In addition, they are influenced by the atmosphere in which the polymerization is carried out [58,60,61].

Fitzer and Müller [62] studied the kinetics of cyclization and oxidation of PAN during heat treatment in air and in nitrogen, using DTA measurements, and observed the cyclization to be a first-order reaction.

Table 7 Kinetic Data of PAN Cyclization in Nitrogen and Air

	Cyclization in N_2			Cyclization in air		
Fiber	E kJ/mol	k_0 sec^{-1}	n	E kJ/mol	k_0 sec^{-1}	n
Dralon T	126 ± 2	1-3 × 10^{11}	0.9 ± 0.1	141 ± 5	3-15 × 10^{11}	–
Dolan	109 ± 2	0.1-3 × 10^{10}	1.0 ± 0.1	111 ± 5	0.1-1.5 × 10^{10}	1.0 ± 0.1

Source: Data from Ref. 62.

The activation energy and the frequency factor were greater in air than in nitrogen (Table 7), indicating that oxygen acts in two different ways. (1) It promotes the initiation of cyclization by forming activated centers and is responsible for their increased number, (2) it inhibits the subsequent cyclization of these centers as indicated by an increase in activation energy.

The difference in behavior of homo- and copolymer PAN, as shown by DTA curves (Fig. 2a,b) in nitrogen and air, suggests [62] that for the copolymer the cyclization is initiated by ions formed by the cleavage in the copolymer, whereas the cyclization of the homopolymer is initiated by radicals.

These polymerization reactions may occur from simple heating of PAN in inert or oxidizing atmosphere, and give rise to a thermally stable cyclized structure which is often referred to as a ladder polymer.

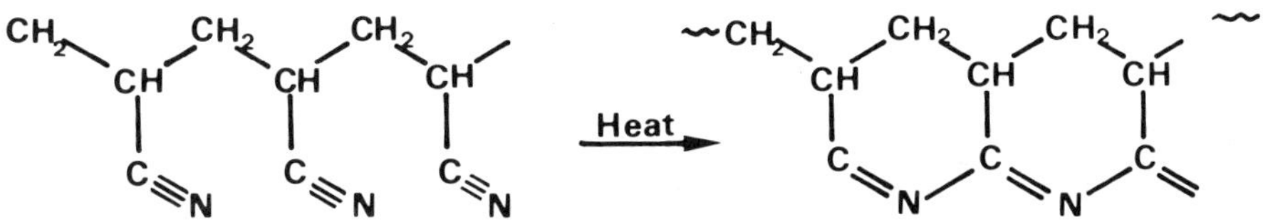

(III)

The chemical formula of the ladder polymer, however, depends upon the atmosphere in which the heat treatment is carried out.

Grassie and McGuchan [58] and Grassie and Hay [60] investigated the heat treatment of PAN in vacuum at 200°C using infrared technique and observed the formation of a ladder polymer which is not dehydrogenated and crosslinked through —C=N—C=N— sequence. They also observed that this polymerization can be initiated by the presence of electrophilic groups such as acids [58].

When PAN is heated in an oxidizing atmosphere, the rate of cyclization is faster and the final carbon product is produced in better yield and with improved mechanical properties (Table 8).

PAN has about 5 percent of comonomer methyl acrylate, which acts as a plasticizer to aid the spinning process. This also facilitates subsequent stretching of the fiber during the oxidation treatment. The fibers are stretched about 2.5 times in a coagulation bath and receive a further stretch of about 14 times in steam at 100°C after washing [63]. This stretching aligns the molecular chains along the fiber axis and helps in raising the Young's modulus. The PAN fibers are spun in tows containing several thousand filaments and having cross-sections which may be circular or shaped like a kidney bean, dog bone, or cauliflower [63], depending upon the method of removing the solvent.

When PAN fibers are heated in controlled oxidizing atmosphere to about 220°C under tension for several hours, several thermally acti-

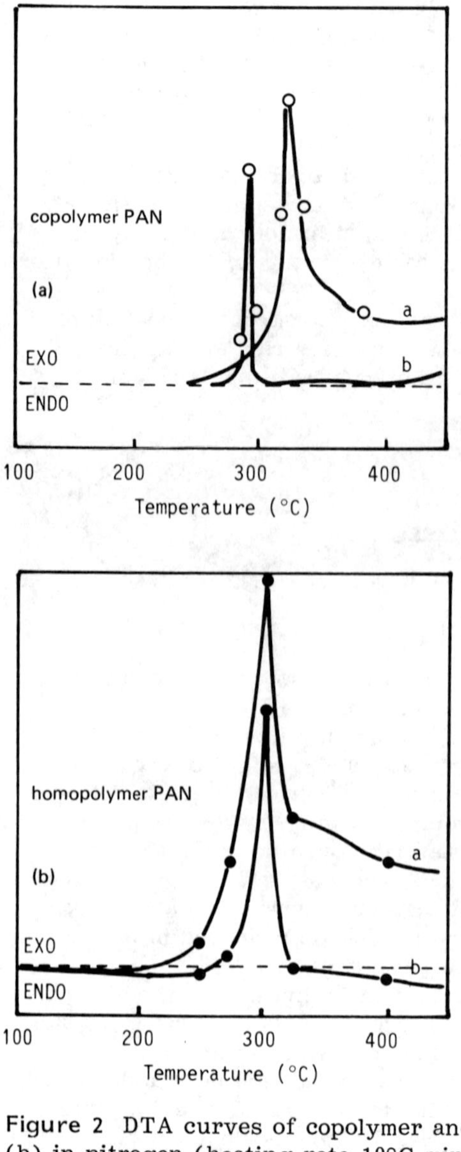

Figure 2 DTA curves of copolymer and homopolymer PAN (a) in air and (b) in nitrogen (heating rate 10°C min^{-1}). (From Ref. 62; photo courtesy of E. Fitzer.)

Table 8 Effect of Preoxidation on Carbon Fiber Yield and Their Mechanical Properties

Temperature of carbonization °C	Loss in weight (%)		Mechanical properties	
	Preoxidized fiber	Unoxidized fiber	Tensile strength (GN m^2)	Young's modulus (GN m^2)
400	15.9	52.2	–	–
600	28.4	59.1	–	–
800	37.8	62.9	1.1	105
1000	43.4	68.0	1.4	161
1500			2.2	238
2000			1.8	300
2500			1.8	384
2800			1.5	426

Source: Mechanical properties: Ref. 56.
Loss in weight: Ref. 73.

vated processes occur [64,65], causing considerable reorganization of the polymer chains and three-dimensional linking of the parallel molecular chains by oxygen bonds. These oxygen crosslinks keep the chain straight and oriented parallel to the fiber axis even after the release of tension following the treatment. The optimum uptake of oxygen in the process is in the range of 8–10% [60,61]. The carbon content of the oxidized PAN fiber changes only slightly. However, the oxidation treatment increases the carbonization yield during any subsequent carbonization process [61].

Raskovic and Marinkovic [66] and Raskovic et al. [67], investigated the stabilization of PAN fibers by oxidation with SO_2 instead of oxygen or air, and observed that considerable amounts of sulfur are incorporated into the fiber chemical structure. This sulfur forms bridges between the adjacent molecular chains, producing a rigid structure. Two temperature zones were recognized. In the low temperature zone (up to 230°C) the reaction proceeds at a low rate and results in a smaller incorporation of the sulfur, which provokes a large decrease in the DTA peak and stops the extension of the fiber under stretching force. This indicates that the fiber has become more stable and oxidation-resistant, so that further high temperature treatment does not degrade it considerably. Above 230°C the reaction is much faster and leads to a certain degradation of the fiber, the extent of which is relatively smaller if a prolonged pretreatment at a lower temperature is carried out. Infrared (IR) studies indicated that the CN groups react slowly at 230°C, but rapidly at 290°C, giving rise to a cyclized aromatized product. Thus, it has been observed that, by combining the low-temperature and high-temperature treatments with SO_2, fibers with larger amounts of crosslinking and initial carbon content can be obtained, which give rise to carbon fibers with better mechanical properties.

1.2.2 Mechanism of PAN Oxidation

Oxidation of PAN fibers has been the subject of numerous investigations, and consequently several different structures have been proposed for the cyclized PAN material. Hantz [68] was the first to put forward a structure for Orlon black fibers obtained by heating in air.

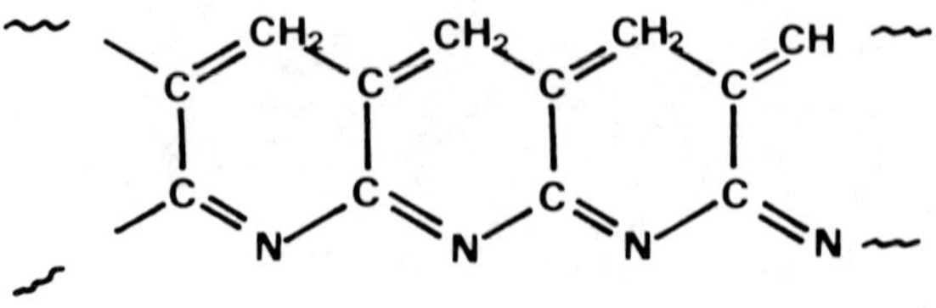

(IV)

The structure was later supported by other investigators [69,70]. However, this structure contains no oxygen and cannot account for the evolution of CO and CO_2 during the carbonization of these oxidized fibers up to 1000°C [71].

Standage and Matkowsky [72], on the basis of IR studies and elemental analysis of oxidized PAN, proposed a different structure with oxygen forming bridging ether links.

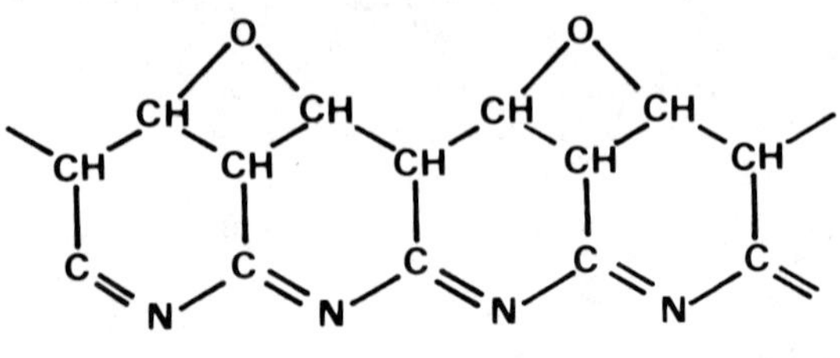

(V)

Watt [73], on the basis of mechanical properties, elemental analysis, IR spectroscopy of PAN fibers after heating in vacuum or in air at 220°C, and evidence provided by model compounds, concluded in favor of the formation of a ladder polymer in which oxygen is present only as carbonyl groups.

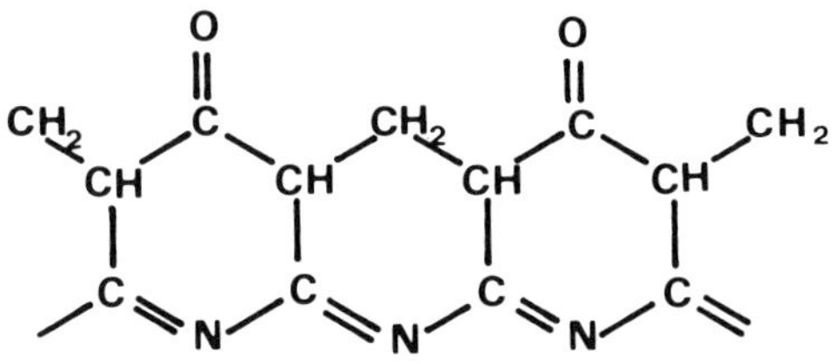

(VI)

This structure was supported by Bahl and Manocha [74]. Kirby et al. [75] and Friedlander et al. [76], after a study of the model compounds, proposed a nitrone-type ladder structure with each nitrogen atom donating its lone pair of electrons to an oxygen.

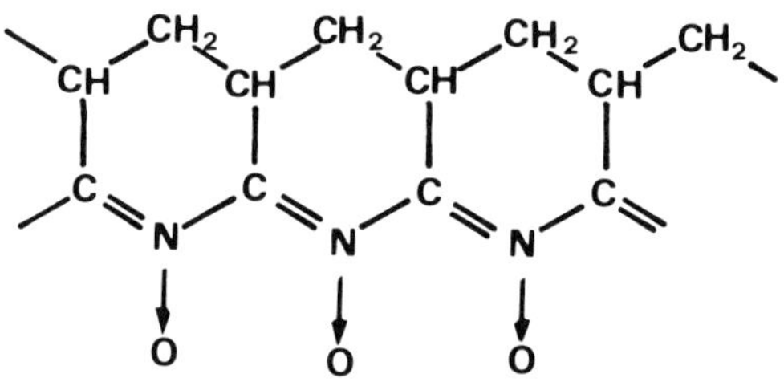

(VII)

However, all these structures (II–IV) have been criticized [77] on the basis of stereochemistry. The carbon atoms in these ladder-type polymer chains are not in one plane, thus leading to a cyclic and not aromatic character. Besides, these structures possess additional polar groups, suggesting that the oxidized PAN fiber should show better mechanical properties than the cyclized PAN fiber, contrary to the observations of Watt [73]. In addition, the nonaromatic character of these structures would suggest a lower thermal stability of preoxidized PAN fibers, although they actually have greater thermal stability.

Strauss and Madorsky [78] and Monahan [79] studied the pyrolysis of high-purity PAN powder and observed the evolution of HCN, which they attributed to an intramolecular reaction followed by chain scission to give other products. But they did not observe the evolution of NH_3. On the other hand, several other workers [80,2,81,82] reported that both HCN and NH_3 were present in the decomposition products. Monahan suggested that NH_3 may be produced by the secondary reaction

$$HCN + H_2O \longrightarrow NH_3 + CO$$

(VIII)

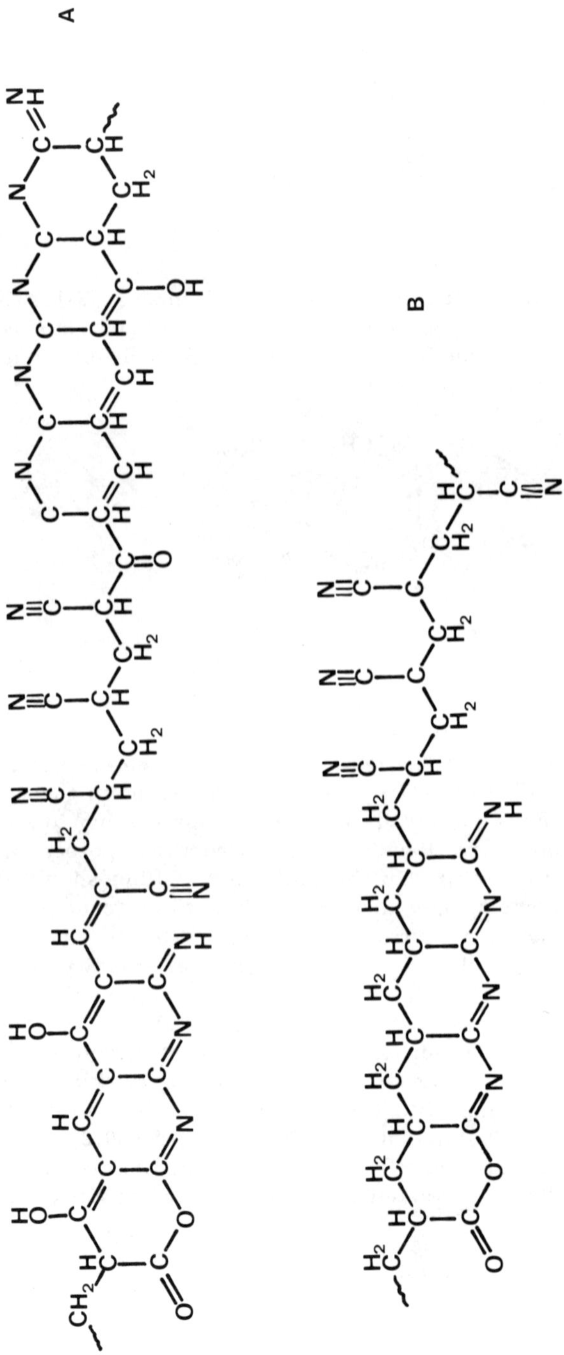

(IX)

whereas Hay [83] and Turner [84] attribute the formation of NH_3 to aromatization of the end groups of a ladder unit, or to the interaction of two terminal groups during the exothermic process which occurs at about 260°C.

The evolution of HCN and water during the oxidation of PAN in air at 160–200°C has been attributed [83] to the oxidation of methylene groups $\leftarrow CH_2 \rightarrow$ to carbonyl groups which then activate the neighboring —CH groups to be oxidized to hyperoxides, and decomposing into HCN and water. The elemental analysis, however, shows that only a fraction [84] of the $—CH_2$ groups is converted into C=O groups which then undergo crosslinking with the remaining $—CH_2$ groups on the adjacent chain [84], and eliminate water. Infrared spectroscopy has shown [85,86] the simultaneous development of a group absorbing at 3330 cm^{-1}, which appears to be a hydroxyl group. In addition, IR spectroscopy provided evidence that carbon double bonds are produced readily on oxidation [87] and add to the stability of the ladder polymer. Thus Goodhew et al. [88] suggested structures A and B (as seen on page 20) as the most likely structures for oxidized and inert heat-treated PAN fibers. These structures are adequate to explain the occurrences in oxidative and inert heat treatments.

The principal structural difference between the two treatments is that while oxidation results in fully aromatic heterocycles carrying hydroxyl and carbonyl groups, the treatment in inert atmosphere produces hydrogenated heterocycles with carbon–nitrogen conjugation. Thus the oxidative treatments produce more thermally stable structures.

It may, however, be pointed out that even this structure cannot account for the evolution of CO_2 during carbonization by heating in inert atmosphere up to 1000°C, as reported by several workers [71,77]. Clearly there is still great uncertainty regarding the reactions which occur during the pyrolysis and oxidation of PAN in air.

Several groups (89–94) have recently carried out Fourier transform IR studies of the initial reactions occurring in the degradation of PAN copolymers at temperatures below the onset of the major exothermic reaction (i.e., $\leqslant$ 200°C) and under reduced pressure of 5×10^{-2} torr. There is no significant weight loss under these conditions and the main reaction involved is the cyclization of acrylonitrile units Fig. 3). The degradation of PAN was found to involve initiation by an anion $x^{\ominus}$, which is presumably derived from an impurity or a degradation product (Fig. 3B). The initiation was followed by cyclization to yield an imine structure (Fig. 3C), which tautomerized into an enamine structure (Fig. 3D), followed by oxidation to give the final pyridone structure (Fig. 3E). The sequences of cyclized groups in the ladder structure impart rigidity along the chain. In addition, the structure of the degraded material during the final pyrolysis to carbon fiber is maintained by extensive interchain bonding between the polymer chains through C=O and N—H groups. The mechanism of the for-

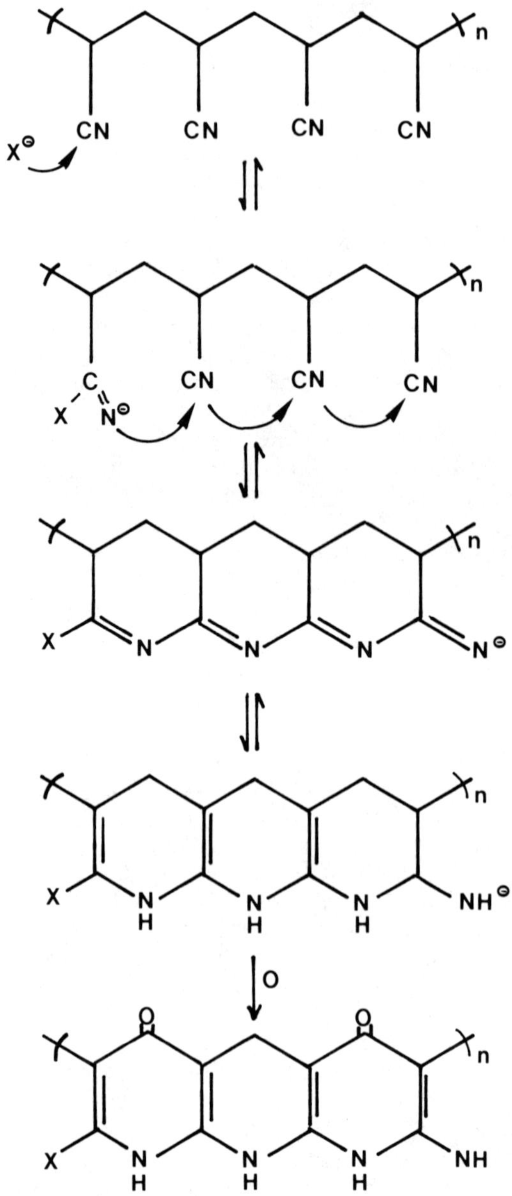

A

B

C

D

E

Figure 3 Decomposition of PAN at 200°C under reduced pressure. (From Ref. 91.)

mation of the ladder structure is, however, different in the presence of copolymers such as methacrylic acid [92], acrylamide [93], and vinyl acetate [94]. These copolymers participate in the initial cyclization reaction and markedly influence the rate of the cyclization process. A comparison of the relative rates of degradation of PAN and copolymers of PAN containing 4% by weight of comonomers, showed that the inclusion of the comonomers, especially methacrylic acid and acrylamide, into the PAN polymer chain considerably enhanced the rate of degradation. The rate of degradation in the presence of vinyl acetate copolymer, although significantly faster than that of PAN homopolymer, was much slower than in the case of other copolymers. A series of studies [157,158] investigated the effect of comonomers in PAN fibers on the stabilization behavior during oxidation in air. These workers measured the length change of laboratory as well as commercially available PAN fibers containing varying amounts of the comonomer (methyl acrylate up to 5%) during isothermal oxidation at 275°C with and without load. Oxidation without load showed initial shrinkage, the extent of which increased with increasing monomer content. The stretching under load, however, increased with increasing monomer content.

1.2.3 Carbonization of PAN Fibers

Having stabilized the stretched structure of the molecular chains, the tows of PAN fibers are subjected to thermal pyrolysis in an inert atmosphere. The noncarbon elements are removed (Fig. 4) as volatiles (i.e., H_2O, HCN, NH_3, CO, CO_2, N_2, and so on [82-84,87,160]) to give carbon fibers with a yield of about 50 percent of the mass of original PAN. The rate of heating in the early stages of the carbonization process is generally low (a few °C min^{-1}) so that the release of the volatiles does not damage the fiber. Most of the volatiles are evolved below 1000°C, and only carbon and some nitrogen ($\simeq$ 6%) are left.

During the early stages of carbonization, in the temperature range 400-500°C, the hydroxyl groups present in the oxidized PAN fibers start crosslinking condensation reactions which help in reorganization and coalescence of the cyclized sections (Fig. 5). This crosslinking probably fixes the structure of the polymer, while the remaining linear segments either become cyclized or undergo chain scission evolving the gaseous products. These cyclized structures undergo dehydrogenation [95] and begin to link up in the lateral direction, producing a graphitelike structure consisting of three hexagons in the lateral direction and bounded by nitrogen atoms [88] (Fig. 5). Watt [96,97] suggested that the lateral molecular growth occurs by condensation reactions in which the carbon atoms in one cyclized sequence fit into spaces left by the nitrogen atoms in an adjacent cyclized sequence. However, Goodhew et al. [88] are of the opinion that, al-

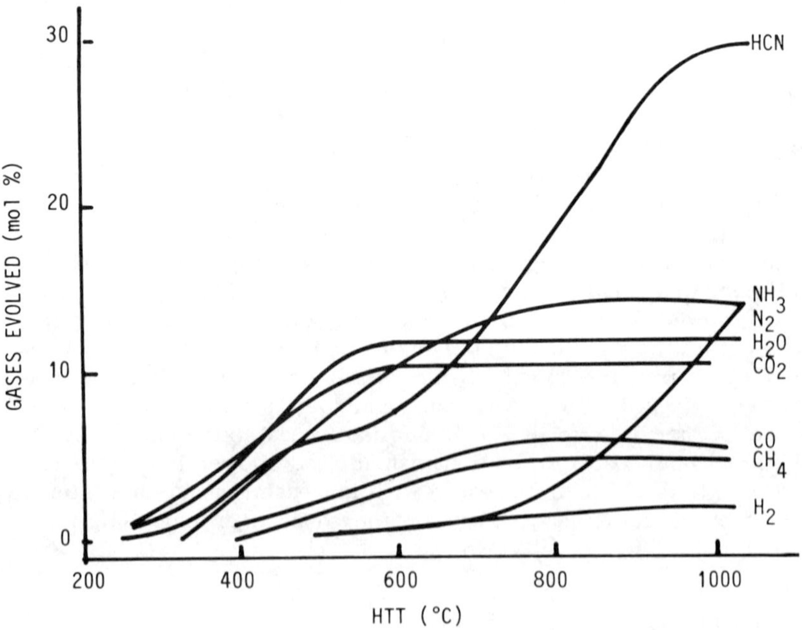

Figure 4 Gases released during pyrolysis of PAN. (From Ref. 160; photo courtesy of E. Fitzer.)

though this is sterically possible, it cannot be the only system causing removal of nitrogen. They argue that a fused heterocyclic ring system requires the presence of 28% nitrogen, although the oxidized system at 700°C contains only 17% nitrogen. Thus, in addition to ring-system carbon, significant amounts of carbon must be available for the interchange mechanism to be operative. The elimination of nitrogen continues above 700°C and reaches a maximum rate at 900°C. At 1300°C the carbon fiber has low nitrogen content. Some authors [88,96] have given values as low as 0.3% although it is generally higher (as much as 2-3%).

Oxidized PAN fibers are capable of producing high-performance carbon fibers in greater yield because they have preferentially an aromatic character which prevents the backbone carbon chain from extensive splitting. In the case of PAN fibers stabilized in inert atmosphere, the aromatization does not readily occur at low temperatures, and therefore it has to be induced at higher temperatures which causes considerable splitting of the main carbon chain and results in a low yield and poor mechanical properties.

The carbon fiber yield can further be enhanced by carbonizing oxidized PAN fibers in an atmosphere of HCl vapors [98] or by

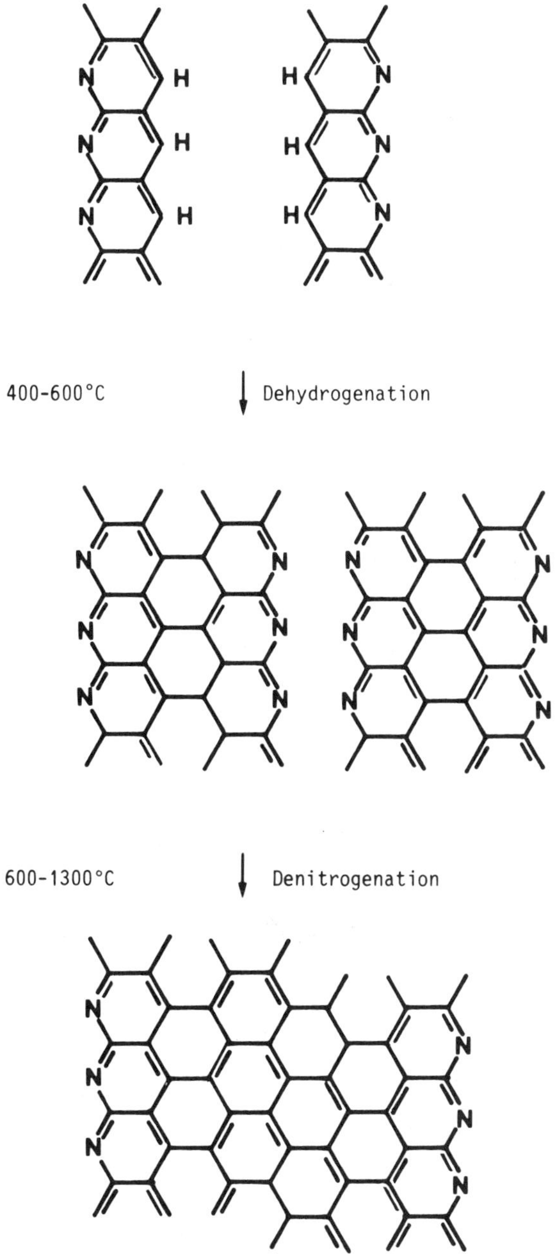

Figure 5 Schematic representation of carbon fiber preparation from PAN. (From Ref. 88.)

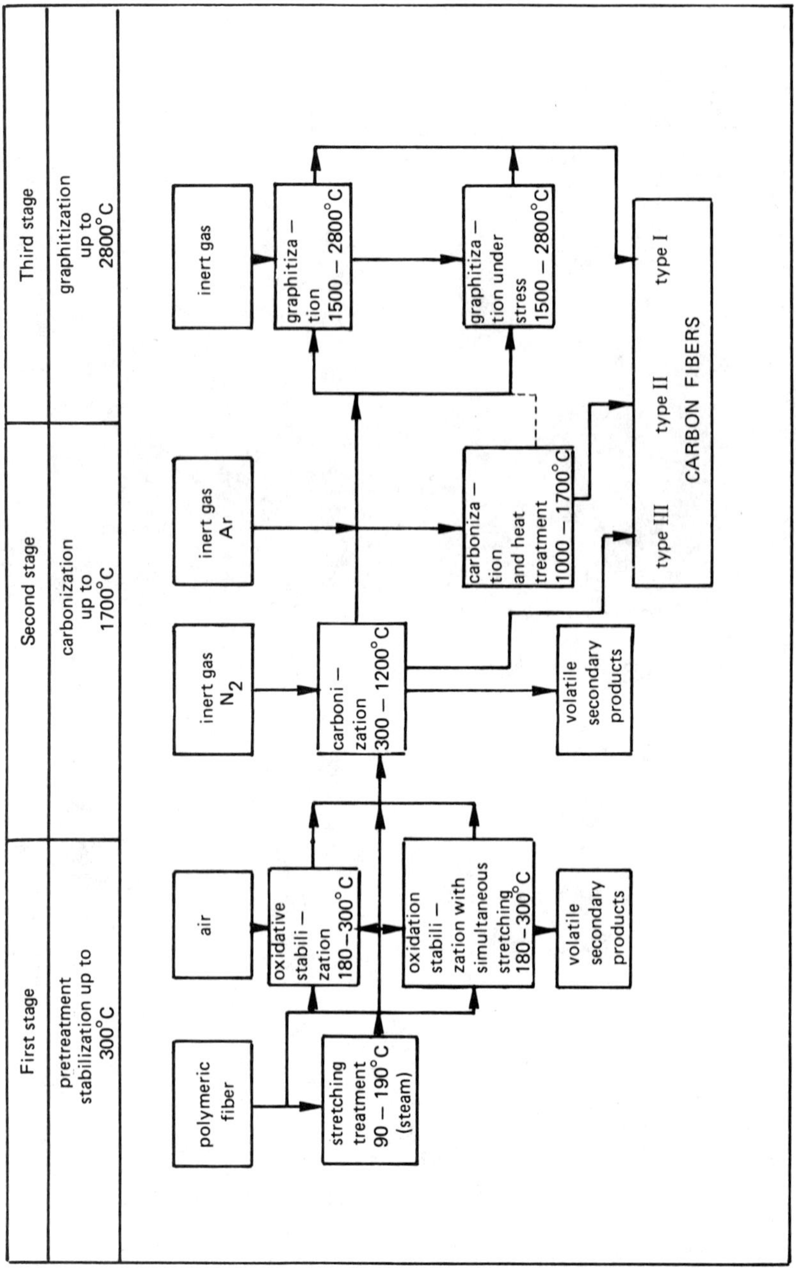

Figure 6 Schematic representation of carbon fiber preparation from PAN. (From Ref. 159.)

spinning the PAN fiber in a clean room [99]. Carbonization in HCl vapors decreases the amount of HCN by eliminating nitrogen as ammonia. It also has a marked dehydrating action and eliminates oxygen as water vapor. Consequently, the yield of carbon increases. Spinning of the PAN fiber in a clean atmosphere results in a decrease in the average number of surface flaws, giving rise to carbon fibers with better mechanical properties.

1.2.4 Graphitization

The carbon fiber obtained after the carbonization process has lost most of its noncarbon impurities and has a graphitelike structure. Further heat treatment at high temperatures even greater than 2500°C does not cause any appreciable decrease in weight of the fiber, but improves the ordering and orientation of the crystallites in the direction of the fiber axis. This results in better mechanical properties of the final carbon fiber. The Young's modulus of the carbon fiber has been found to be directly related to the final heat treatment temperature during graphitization.

1.2.5 Hot Stretching of PAN Carbon Fibers

The Young's modulus of a carbon fiber depends upon the preferred parallel alignment of the basal planes of the small turbostatic crystallites with the fiber axis. Thus, the Young's modulus can be improved by increasing this orientation. Bacon and Schalamon [100] have shown that high-modulus carbon fibers could be prepared from rayon by hot stretching of the carbon fibers during the graphitization treatment, and observed that the modulus may be controlled by the amount of stretch. Based on these observations, Johnson [101] stretched loops of PAN carbon fibers by amounts up to 28.6 at 2800°C and obtained carbon fibers with modulus improved from 522 to 689 GN m^2. The strength of the carbon fiber was also found to be increased by hot stretching.

A complete flow diagram for the preparation of carbon fibers from PAN is given in Fig. 6.

1.3 CARBON FIBERS FROM RAYON

Cellulose is the second important raw material for large-scale production of carbon fibers because it degrades without melting. Natural cellulose fibers such as those of cotton and ramie are not preferred for the production of carbon fibers because they produce discontinuous filaments, have a low degree of orientation, and are associated with impurity materials such as lignin. They have a high degree of crystallinity and, therefore, result in low yields of carbon fibers with poor mechanical properties. Accordingly, they are considered inferior to synthetic cellulose fibers.

The most commonly employed cellulosic fibers for producing carbon fibers are textile-grade rayon, regenerated cellulose, and cuprammonium rayon, although polynosic rayon [102] and viscose rayon fibers [103] have also been carbonized.

The formation of carbon fibers from rayon fibers also involves the usual three stages: low temperature decomposition (< 400°C), carbonization (< 1500°C), and graphitization (> 2500°C). The yield of carbon fibers obtained by the pyrolysis of rayon is usually very low (varying between 10 and 30%) and depends upon the nature of the precursor fiber, the rate of heating, and the environment (atmosphere in which the pyrolysis is carried out). The molecular weight of the cellulosic precursor also plays a vital role as it determines the number of end groups available for initiation of the degradation process. A slow rate of heating results in a higher yield but it is less economical, and therefore, pyrolysis is carried out in a reactive atmosphere [11,104-106] or in the presence of flame retardants [12]. These chemical species change the course of the decomposition process so that it occurs at a lower temperature but at a much faster rate and results in a greater yield of the carbon fiber [107]. Since cellulosic materials suffer disruption of each ring-to-ring linkage and extensive weight loss, the graphitization is usually carried out under stress orientation so that the atoms can be juxtaposed to fit into the graphite crystal.

1.3.1 Low-Temperature Degradation

The repeat unit of cellulose in rayon

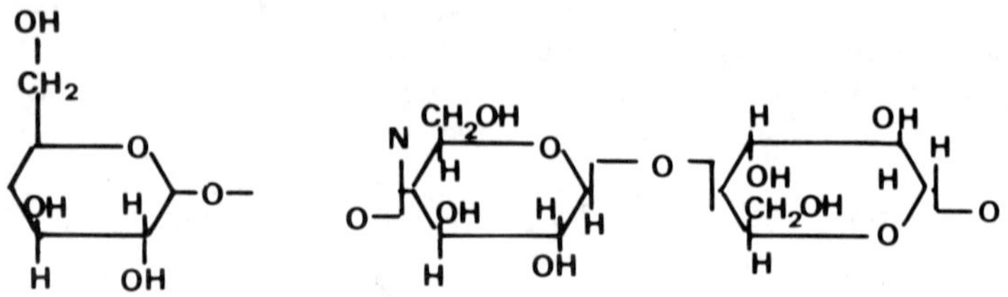

(X)

contains hydrogen and oxygen, which can result in the elimination of five molecules of water giving a weight loss of 55.5%. The actual weight loss is of the order of 70-90%. This excessive weight loss has been attributed to chain-splitting reactions involving CO_2, CO, alcohols, ketones, and a number of other carbon-containing compounds [106,108–112]. In addition, the intermediate dehydrated stage produces levoglucosan when the pyrolysis is carried out in an inert atmosphere. This levoglucosan decomposes to give volatile carbon products during the later stages of the pyrolysis process, resulting in a further loss of carbon.

Although pyrolysis of cellulosic materials has been studied extensively, and a number of decomposition products have been identified, the detailed mechanism for the complete conversion of cellulose into carbon was first presented by Bacon and Tang [113]. These workers studied the pyrolysis of regenerated cellulose fibers and filaments up to 500°C and used IR absorption spectroscopy, thermogravimetric analysis, and gas evolution data. According to them, although the pyrolysis process involves four different stages, the major pyrolytic decomposition takes place between 210 and 320°C with an ultimate weight loss of about 90%. The weight loss, however, varied between 70-90% depending upon the raw material and the rate of heating. The rate of heating was usually kept low to avoid excessive tar formation and tar deposition onto the filaments, which results in brittleness of the carbon fiber and disruption of the filament structure.

Stage I (25-150°C) is essentially a reversible process involving elimination of absorbed water amounting to about 12%. In addition, a small increase in the degree of lateral order has been observed [114] during this stage. This is followed by the splitting up of the structural water in stage II (150-240°C) from hydrogen and hydroxyl fragments present, preferably in equatorial positions. Infrared analysis showed the formation of C=O and C=C bonds, indicating that the dehydration process is essentially intramolecular.

The major part of the thermal degradation results from thermal cleavage of C—O and C—C linkages above 240°C (stage III 240-400°C), and leads to the formation of large amounts of tar, H_2O, CO, and CO_2. Each cellulose unit breaks down into a residue containing four carbon atoms which then repolymerize into a carbon polymer ultimately producing a graphitelike structure above 400°C (stage IV) by condensation reactions involving the removal of hydrogen.

The possible reaction scheme [113] for the various stages of the polymerization process is reproduced in Figure 7. A yield of a four carbon residue from each cellulose ring unit corresponds to an ultimate weight loss of 70.4%, which is actually obtained under certain conditions. The CO_2 to CO ratio in the gaseous products and amounts of oxygen found in the high-temperature residue fiber indicate that some C=O containing species participate in the polymerization process before evolving as CO.

Degradation of Rayon in Reactive Atmosphere

The carbon fiber yield and the processing rate can be markedly enhanced by carrying out the low-temperature pyrolysis of rayon in the presence of a reactive atmosphere, such as air or oxygen [115–117],

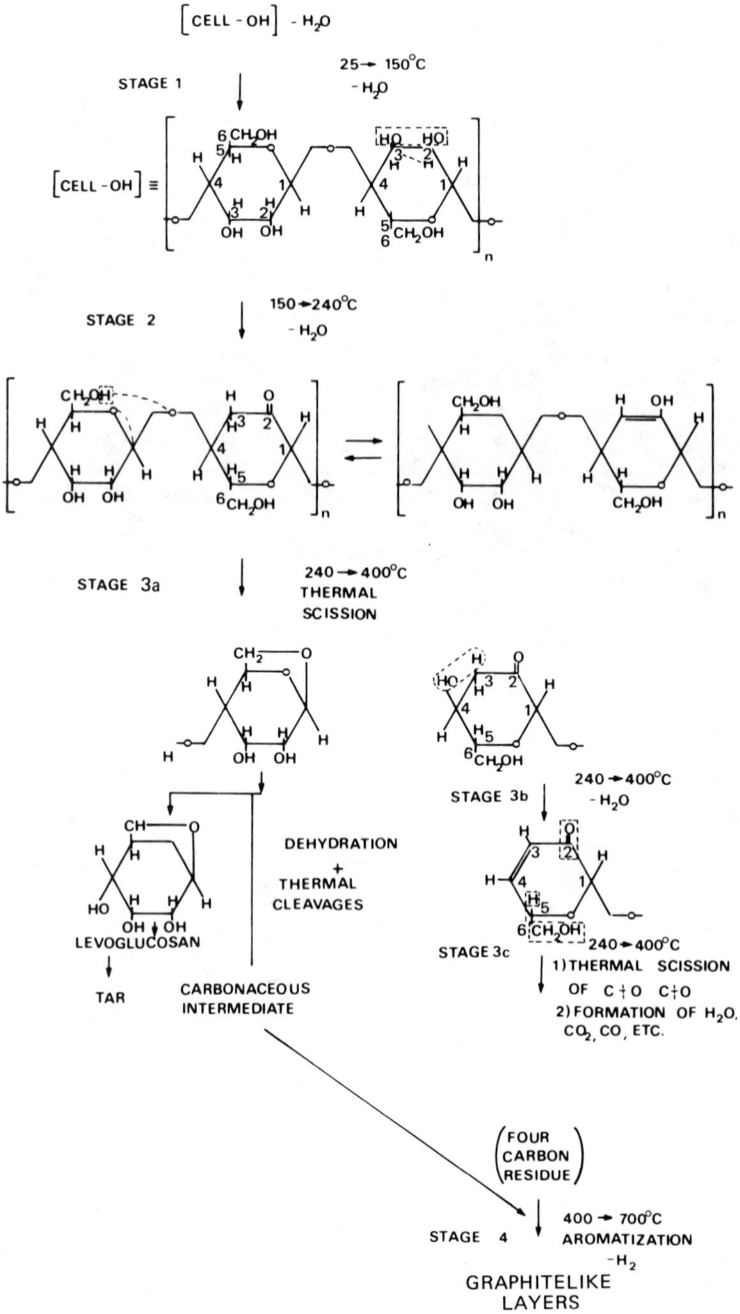

Figure 7 Reactions involved in the conversion of cellulose into carbon fibers. (From Ref. 113.)

chlorine [118,13,111], and HCl vapor [11,119]. The presence of a reactive atmosphere promotes dehydration of cellulose and inhibits the formation of tars [11,106,111]. It is evident (Fig. 7) that the formation of levoglucosan, which later produces tar, occurs through primary hydroxyl groups. Pyrolysis in the presence of oxygen [106, 120–123] results in the oxidation of C—6 methylol groups into carboxylic groups and thus prevents the formation of levoglucosan. In addition, the carboxylic acid groups, because of hydrogen bonding, can give rise to a highly crosslinked intermediate, resulting in a better carbon yield. The dehydration starts much earlier but continues through a wider temperature range.

Pyrolysis studies in HCl vapor using weight loss measurements, x-ray fiber diffraction, and IR analysis of the residual species [11] showed that the pyrolysis started as early as 110°C and the weight of the fiber decreased in a stepwise manner mainly through the elimination of water. The ultimate yield of the fiber corresponded to a yield of five carbon atoms from each cellulose ring unit (Fig. 8). The influence of reactive atmosphere on the low-temperature pyrolysis is shown in Figures 9 and 10.

The processing time and the carbon yield vary with the nature of the reactive atmosphere. The time of processing has been found to vary between 20 minutes and 10 hours, and the carbon yield for rayon carbonized up to 1000°C varied from 23% in air or oxygen atmospheres [115], to 27% for a chlorine atmosphere [13] and to 35% in HCl vapor [11].

Degradation of Rayon in the Presence of Flame Retardants

The efficiency of a reactive atmosphere in increasing processing rate and improving carbon yield, however, is limited by the rate of diffusion of the reactive gaseous species into the interior of the fiber, which is a slow process and may require several hours. The rate of diffusion can be enhanced by increasing the temperature, but this may lead to excessive degradation of the fiber structure and is therefore not resorted to. However, this diffusion limitation has been overcome by impregnating rayon fibers with certain reactive chemicals from aqueous solution before the heat treatment. These substances, which include nitrogenous salts [124] and certain phosphoric esters [12,125], are effective flame retardants [126,127] and alter the course of the carbonization process. The decomposition of cellulose in the presence of these flame retardants occurs at a lower temperature, at a faster rate, and the amount of carbon formed is larger than that formed without the flame retardant. The flame retardant either catalyzes a reaction involving elimination of hydroxyl groups or reacts with the hydroxyl group and promotes dehydration of the cellulose material at low temperature [128]. The pyrolysis of rayon in their

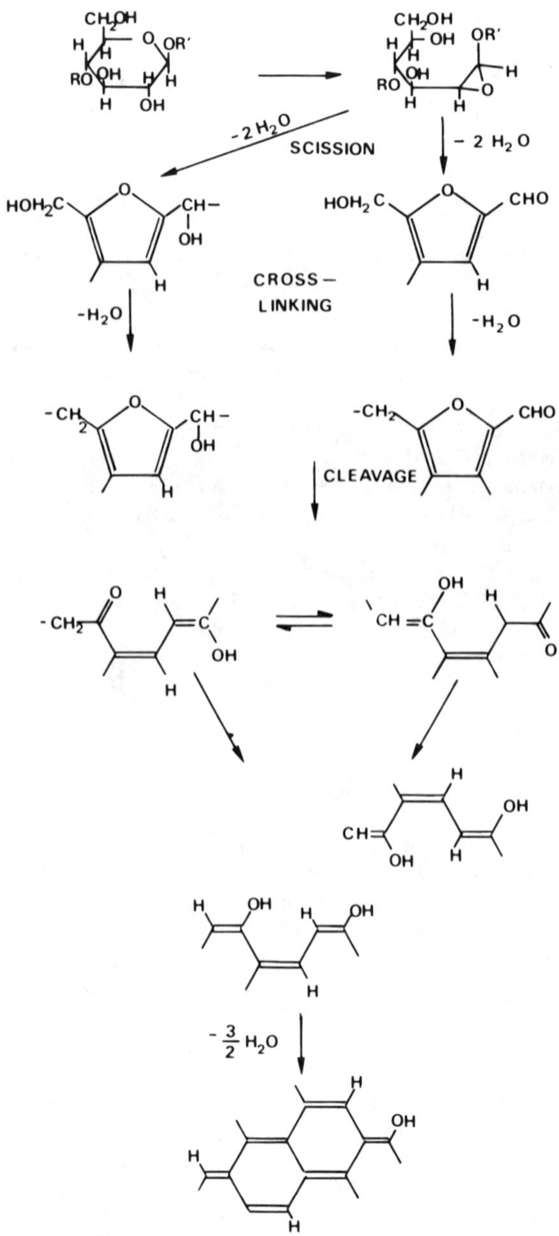

Figure 8 Conversion of cellulose into carbon in the presence of hydrogen chloride. (From Ref. 11.)

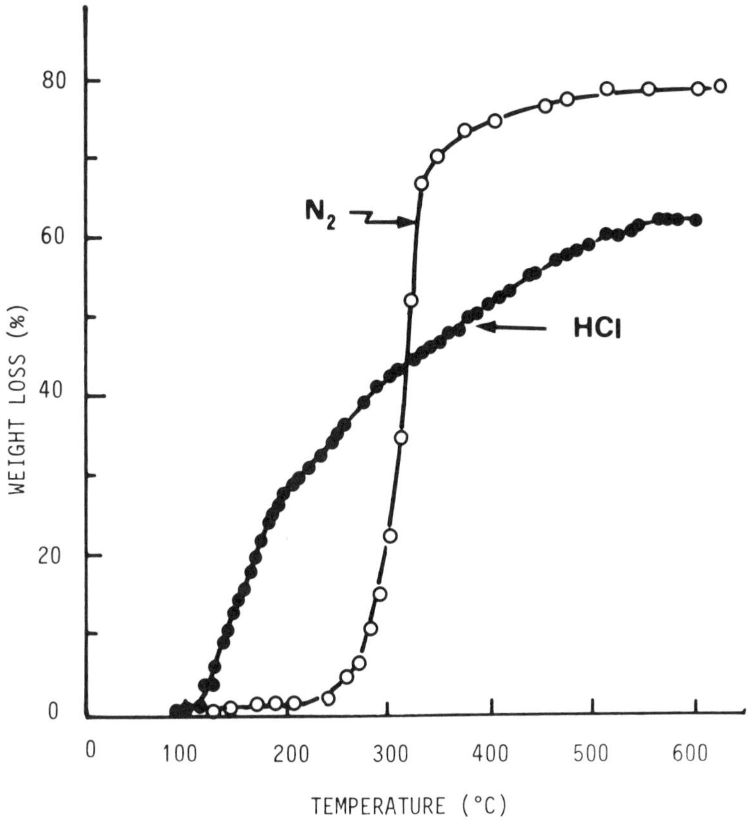

Figure 9 Effect of chlorine atmosphere on weight loss during pyrolysis of rayon. (From Ref. 132.)

presence produces dehydrocellulose exclusive of any levoglucosan. The weight loss up to 200°C was found to be exclusively due to the volatilization of water, and no carbon-containing gases were detected. Consequently, a greater yield of carbon was obtained.

Pyrolysis up to 800°C of rayon treated with phosphoric compounds such as bis(chlorodiphenyl phosphine)decarborane (BCDPD), and dimeric chlorophospha(III)—O—carborane (DCPC) gave a better carbon yield and a substantial improvement in oxidative stability at 250°C [12].

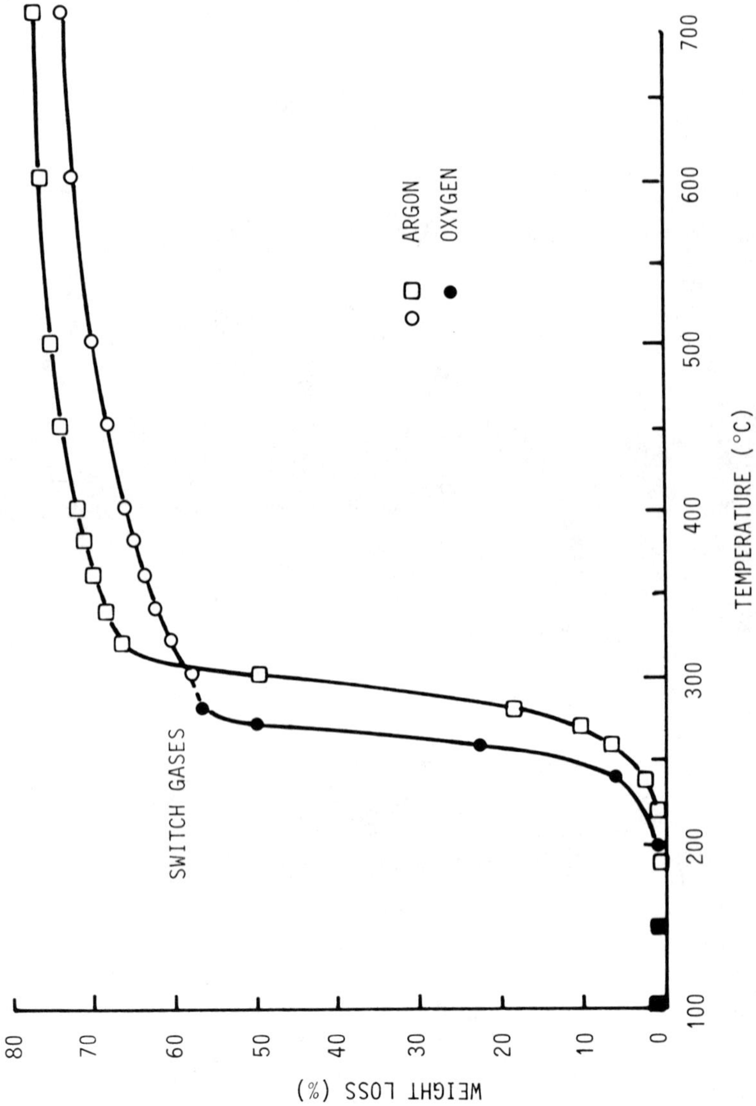

Figure 10 Effect of oxygen atmosphere on weight loss during pyrolysis of rayon. (From Ref. 132.)

1.3.2 Carbonization at High Temperatures

Infrared spectra of extracts of chars obtained from cellulose at temperatures between 240 and 400°C reflected only minor changes in C=O and essentially no changes in C=C bond [113], showing that the C=O and C=C conjugated system is intact. However, IR spectra above 400°C showed very weak C—H and C=O bonds. The C=O bonds diminished still further as the temperature was increased. This indicates that around 400°C the rayon fiber suffers major molecular breakdown due to the thermal scission reactions of C=O bonds, so that the rayon fiber is ready to be carbonized to higher temperatures ranging between 1000 and 1500°C. This carbonization is usually carried out in inert atmosphere, although the use of reactive atmospheres has been suggested in the literature [11,111,115–119]. Usually, the carbonization is carried out in a matter of seconds.

Carbonization at higher temperatures has also been carried out under tension [132], which improves the mechanical properties. However, it has been found that the mechanical properties can be improved further if the tension is applied during the graphitization process. A marked reduction in resistivity has been observed [122] during heat treatment between 500 and 1000°C, indicating the formation of carbon atoms with free unattached valencies [129]. The carbon residues produced as a result of these carbonization reactions are suggested to be furan derivatives (XI) (XII).

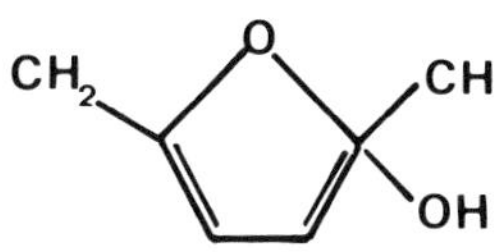

(XI)

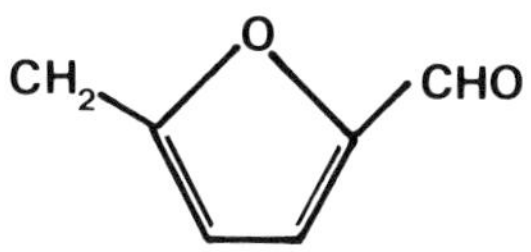

(XII)

These carbon residues crosslink with increasing temperature and result in the formation of aromatic ring structures by the elimination of hydrogen.

1.3.3 Formation of Graphitic Structure from the Four-Carbon Residue

X-ray analysis of regenerated cellulose filaments [130] with various degrees of molecular orientation, before and after carbonization showed that a highly oriented rayon fiber produced a highly oriented carbon fiber. This indicates that the graphitized layers are formed with an orientation in the direction of the original cellulose chain. Thus, the four-carbon residues (furan derivatives) align themselves with their long dimensions parallel to the b axis, joining together to

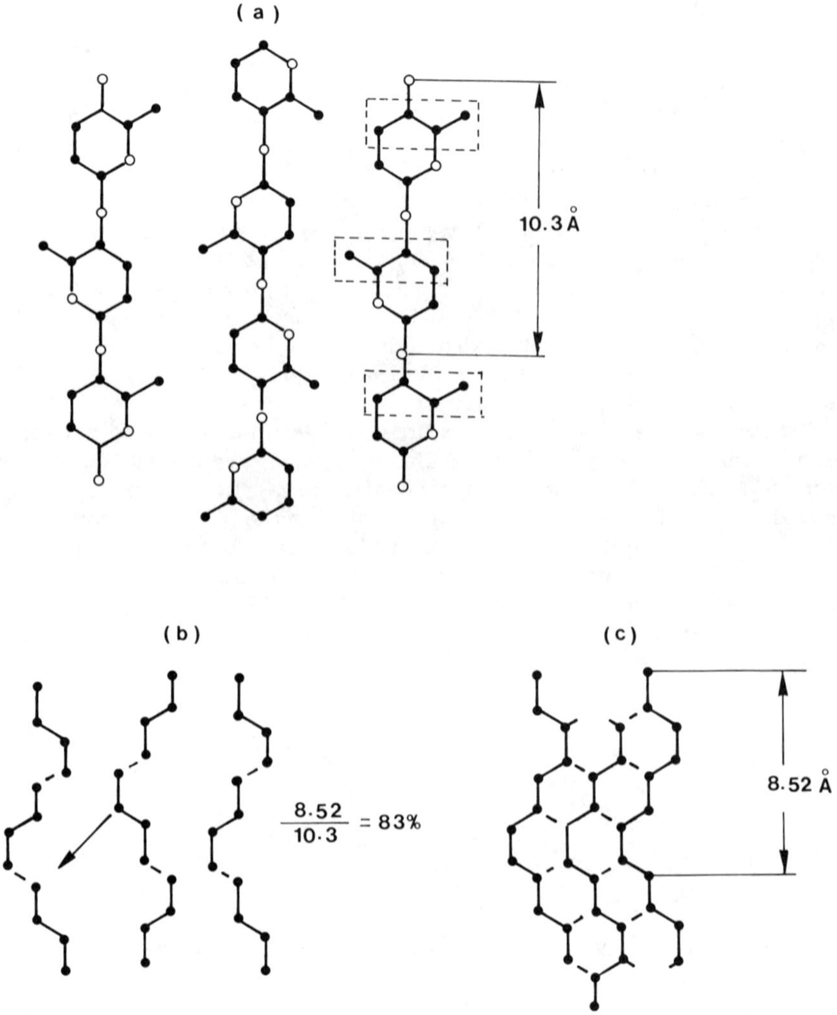

Figure 11 Longitudinal polymerization of cellulose ring unit residues into graphitic structure. (From Ref. 130; photo courtesy of R. Bacon.)

form a chain polymer along the original cellulose direction. Adjacent chains then condense to form six-carbon graphitic structures. This proposed longitudinal polymerization process is illustrated in Figure 11.

According to this scheme the cellulose b axis, which is 10.3 Å long, must shrink to twice the length of the graphite b axis (8.25 Å), resulting in a length yield of about 83%. In actual experiments a length yield of 76% was observed for highly oriented filaments. The difference between the observed and the theoretical length yield has been attributed to several alternate reactions, such as evolution of tar during the low-temperature carbonization, which results in shortening of the chain. Ruland [131], however, observed that the original orientation texture of the fiber is almost completely lost during the carbonization process at 240°C, and remains low until 900°C, and that the cellulose fiber with the highest initial orientation produced a carbon fiber with the lower final orientation. Thus, the ultimate orientation cannot be adequate unless graphitization is carried out under stress.

Graphitization is usually carried out above 2800°C for a fraction of a second under stress [132]. The Young's modulus of the resulting graphitized fiber is directly related to the graphitization temperature and the amount of stretch applied during this stage.

1.4 PREPARATION OF CARBON FIBERS FROM PITCH

Pitch is an important and low-cost material for the manufacture of carbon fibers. Several types of pitch sources have been attempted but most of them are part of the patented literature [133—136]. The most commonly used pitches are those obtained from petroleum asphalt, coal tar, and PVC [137].

Pitches are thermoplastic in nature. Therefore, it is difficult to carbonize them without some kind of thermosetting to maintain the filament shape during pyrolysis. Thus, ordinarily pitches are extruded into the fibrous form at some stage of their thermal treatment, then thermoset between 250 and 400°C, and subsequently carbonized and graphitized between 1000 and 2500°C. The flow diagram of the process is shown in Figure 12.

The suitability of pitch for the production of carbon fibers is determined by its ability to spin and be converted to a nonfusible state so that the fibers can be carbonized without melting. These properties are largely dependent upon their chemical composition and their molecular weight distribution. Thus, petroleum and coal tar pitches are generally given a pretreatment, although PVC pitches can be used as such provided their thermal degradation is carefully controlled.

The molecular weight of the pitch is an important factor which determines the temperature and the speed of spinning by controlling the viscosity of the melt and the melting range. This can be controlled by the addition of certain substances [138,139], by extraction with certain solvents [140], or by distillation [141]. If the molecular weight is too low, the condensation reactions can be enhanced by the addition of tetramethyl thiuram [138]. However, if the molecular

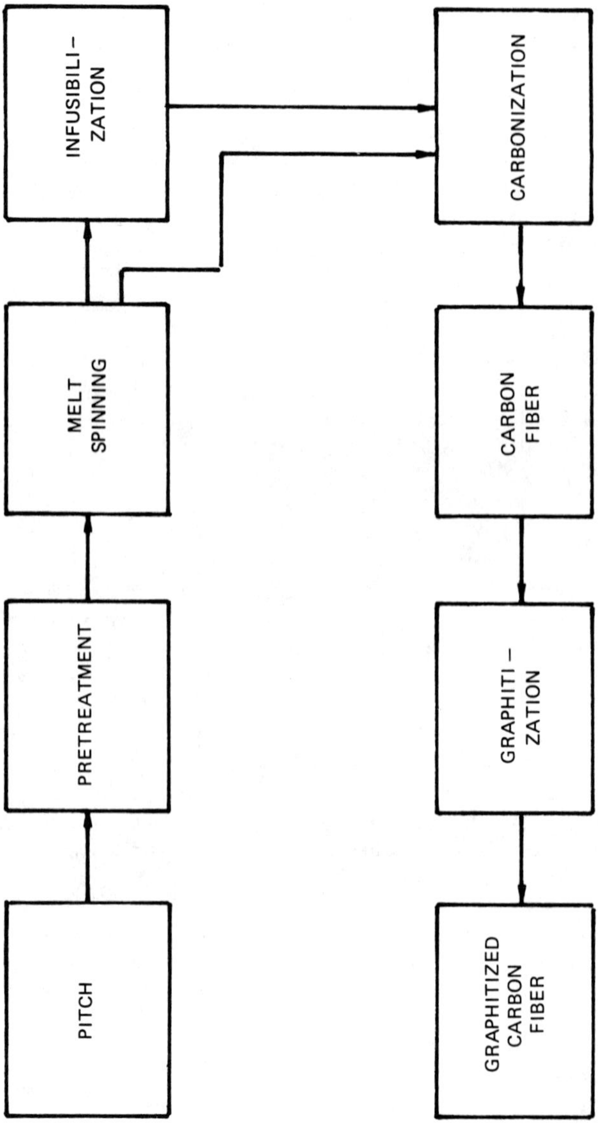

Figure 12 Flow diagram for pitch-based carbon fiber process. (From Ref. 15.)

weight is too high and renders the melt too viscous, the viscosity can be decreased by the addition of certain plasticizers [139].

Pitches are made up of numerous organic compounds with aggregates of condensed benzene-ring systems separated by and carrying alkyl chains. The composition of a pitch, therefore, depends upon its source. Petroleum and coal tar pitches also contain appreciable amounts of sulfur (3-4 atoms per molecule) [138], but the type of bonding involved is not precisely known [138].

PVC pitch is also a mixture of several kinds of molecular species consisting of polynuclear aromatic compounds possessing three or four aromatic nuclei composed of three or four rings per molecule. On the basis of physical analysis using infrared (IR) or ultraviolet (UV) technique, PVC pitch has been suggested to have the following structure with the formula $C_{62}H_{52}$ [15] (XIII).

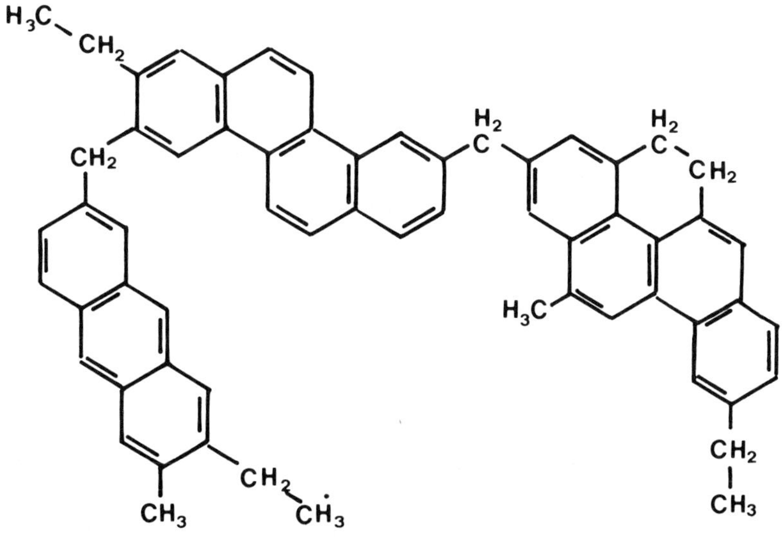

(XIII)

The pitch was produced by the thermal decomposition of PVC at 400°C under nitrogen for 30 minutes in 29% yield. It has a melting range between 150 and 200°C and a molecular weight between 700 and 800.

1.4.1 Carbon Fibers from PVC Pitches

The first pitch carbon fibers were prepared by Otani [15] by the pyrolysis of PVC pitch fibers. The pitch fibers were obtained by extruding molten pitch through a pinhole on the bottom of the cell used for melting pitch. The pitch filaments were extended and solidified as they flowed down through air, and were wound onto bobbins at high speed.

The carbonization of pitch fibers involved, as usual, an oxidation pretreatment in ozonized air below 70°C, and a subsequent heat treatment in air at a rate of 1.5°C min^{-1} up to 260°C. The final temperature was maintained for 3 hours. The ozone treatment creates a kind of intermolecular crosslinking through the formation of an ozonide or a peroxide type of linkage by the addition of ozone to the aromatic nucleus [16]. In addition, some ring scission reactions also occur, giving rise to the formation of aldehydes and ketones. These crosslinking reactions render the filament completely insoluble in all solvents and infusible on further heat treatment.

Goodhew et al. [88] are of the view that oxidation of the fiber occurs when heated in air between 200 and 300°C. The oxygen diffuses through the sheath and reacts with the α-hydrogen atoms, resulting in the formation of hydroxyl and acid groups which cause condensation reactions at higher temperatures producing crosslinks and causing infusibility. After the oxidation treatment the sheath and the core, respectively, contain about 30% and 22% by weight of oxygen [16,88].

The treatment with ozone, although it can be replaced by a slow treatment in air, produces crosslinking at a comparatively low temperature, resulting in the formation of the insoluble sheath at the lower temperature and facilitating the subsequent treatment. The treatment in air has to be carried out at a higher temperature to produce crosslinking and most of the aromatic nuclei remain undegraded [24]. Oxidizing agents other than oxygen, such as nitrogen dioxide [135], peroxides [142], and chlorine [134] have also been suggested for the preoxidation treatment of the PVC pitch fiber. It has also been suggested that the addition of materials such as benzene sulfonyl chloride may increase the rate of oxidation without any loss of properties of the resulting fibers [143].

Thermally stable fibers obtained after the preoxidation treatment were, then, carbonized in a stream of nitrogen by heating at a rate of 5°C min^{-1} up to a final temperature varying between 1000 and 1350°C. The final temperature was maintained for 15 minutes. This carbonization process resulted in an increase in strength, electrical conductivity and specific gravity, but a decrease in the H/C ratio. The total weight loss during preoxidation (increase in weight) and carbonization (decrease of 29.8% of preoxidized weight) was $\simeq$24% of the original pitch filament.

This process for the preparation of carbon fibers has been modified into a much simpler one by Araki and Gomi [144]. According to their procedure, specially prepared pitch material is charged into a centrifugal spinning machine to spin fine continuous pitch multifilaments, which are made infusible by treatment in hot air and subsequently rapidly baked at temperatures up to 1000°C under nitrogen atmosphere to obtain carbon fibers.

However, the carbon fibers obtained either by Otani or by Araki and Gomi were isotropic and had poor mechanical properties. Otani et al. [145] studied the effect of applied tensile load during the heat treatment of PVC pitch carbon fibers on their mechanical properties and observed that a considerable improvement in mechanical properties could be effected by high-temperature hot stretching. It has also been observed [146–148] that the carbon fibers produced from a number of pitch materials were isotropic and amorphous. These investigators, however, could produce high-modulus, high-strength graphite fibers when these carbon fibers were strained at temperatures in excess of 2000°C. Fiber strength of 2.18 GN m^2 and Young's modulus of 440 GN m^2 have been achieved.

1.4.2 Mesophase Pitch Carbon Fibers

The hot-stretching process for the production of pitch carbon fibers is a difficult and costly process. Therefore, attempts were made to produce oriented carbon fibers from other varieties of pitch materials. One such approach employed special pitches consisting of aromatic molecules with structures capable of forming anisotropic pitch fibers. These pitches were prepared [149] by thermal treatment of small polynuclear aromatics such as phenanthrene, benzopyrene, and chrysene or were obtained from the byproducts of the thermal cracking of petroleum fractions or of certain crudes above 700°C, preferably between 1000 and 1200°C.

Barr et al. [7] pointed out that ordinary commercial pitches are mixtures of numerous aromatic compounds with average molecular weights between 400 and 600. When these pitches are heated for prolonged periods of time above 350°C, the molecules undergo a succession of dehydrogenative condensation reactions, forming large molecules which aggregate into a liquid crystalline phase with a nematic order. This phase, called the mesophase, has a higher surface tension than the low molecular-weight isotropic liquid phase from which it grows. It forms small liquid droplets which grow in size, coalesce into larger spheres, and eventually become extended anisotropic regions.

The amount of mesophase continues to increase, and when it reaches about 45%, a phase inversion occurs and the anisotropic phase becomes continuous and surrounds the droplets of the remaining isotropic material. With increasing heat treatment temperature the mesophase polymerizes and solidifies into coke. This solidification must be avoided in the production of carbon fibers.

Singer [6,150] observed that fibers with a high degree of preferred orientation could be spun from mesophase pitches containing a liquid crystalline phase, and that these fibers could be easily converted into high-strength and high-modulus graphite fibers with a well-ordered graphitic structure. A similar approach has been used by Japanese workers [151] to produce well-oriented carbon fibers.

While measuring the viscosity of acenaphthalene pitch between 450 and 500°C, Singer [152] observed that this "taffylike" stage existed for a considerable time before the material solidified, so that it was possible to draw fibers from the mesophase. Thus, the first mesophase pitch fibers were drawn from acenaphthalene pitch at 450°C by repeatedly dipping a quartz rod into the pitch as it gradually thickened. This success gave rise to the development of spinning equipment and also resulted in attempts to use low-temperature melting pitches which could be produced in a separate operation and then spun at a temperature well below the rapid coking temperature [153, 154].

Spinning of Mesophase Pitch: Mesophase pitches can be spun into fiber form by using spinning modes such as centrifugal spinning, jet spinning, or conventional melt spinning. Continuous multifilament mesophase yarn has been obtained by using commercial melt spinning equipment [24]. The spinning is done through an orifice with 0.3 mm diameter. Thus, for spinning 10 μm diameter filaments at 2.5 m/sec, a draw ratio of about 1000:1 was required. By careful control of the various spinning parameters, the filament diameter can be reduced to as low as 7.8 μm by increasing the wind-up rate [7].

Stabilization of Mesophase Fibers: Mesophase pitches are also thermoplastic and must be rendered infusible by a suitable oxidation treatment in order to preserve their preferred orientation during the carbonization process. As these pitches have higher softening points, the oxidation treatment can be carried out at a higher temperature, where diffusion and oxidative polymerization reactions are faster. Consequently, stabilization against melting takes only a few minutes as compared to hours needed for stabilizing conventional pitch fibers.

The stabilized pitch fibers are ready for the carbonization and graphitization process at any temperature, without need of high-temperature stretching.

Mesophase pitch fibers have a high degree of preferred orientation, and retain this original orientation during the thermosetting step. The degree of preferred orientation increases with heat treatment temperatures above 1000°C. After heat treatment at 3000°C, the Young's modulus approaches a value of 886 GPa, close to the graphite single-crystal value of 1000 GPa [7].

Union Carbide produces pitch fibers in two forms: as fiber mat and as continuous filament yarn. The two varieties are primarily differentiated by the spinning technique, but differ significantly in appearance, properties, and cost potential.

Carbon fiber mat made from pitch designated as Thornel type P mat was marketed in 1973 and its cost at present is about $17.60/kg. However, the projected selling price at a multimillion kg/year level is

Table 9 Comparison of Properties of Carbon Fibers from Different Precursors

Precursor	Density ($g\ cm^3$)	Young's modulus (GPa)	Electrical resistivity (10^{-4} ohm · cm)
Rayon[a]	1.66	390	10
Polyacrylonitrile[b] (PAN)	1.74	230	18
Pitch (Kureha)			
LT[c]	1.6	41	100
HT[d]	1.6	41	50
Mesophase pitch[e]			
LT	2.1	340	9
HT	2.2	690	1.8
Single crystal[f] graphite	2.25	1000	0.40

[a]Union Carbide, Thornel 50.
[b]Union Carbide, Thornel 300.
[c]LT, low-temperature heat-treated.
[d]HT, high-temperature heat-treated.
[e]Union Carbide type P fibers.
[f]Modulus and resistivity are in-plane values.
Source: Data from Ref. 152.

expected to be much lower. Scanning electron micrographs of the product [155] showed that the filaments had a random, layered orientation and an average diameter of about 10 μm [155]. The curved nature of the filament resulted from the spinning process.

The pitch-based continuous carbon fibers (Thornel type P continuous yarn) are round and have a diameter of 7.8 μm and a density of 1.95 $g\ cm^3$. They have mechanical properties comparable to PAN or rayon carbon fibers, their strength being somewhat lower (Table 9).

1.5 CARBON FIBERS FROM PHENOLIC PRECURSORS

Kuwamura and Jenkins [50] prepared glassy carbon fibers by the pyrolysis of a phenol-hexamine polymer. The polymer was extracted at 120-135°C by forcing through an orifice under pressure of nitrogen. The fiber was drawn in air by wrapping it around a rotating frame. The texture and diameter of the fiber are determined by the speed of rotation of the frame.

The fibers were obtained through carbonization, by heating slowly in nitrogen at the rate 1°C min^{-1} to 350°C, followed by rapid heating to a temperature of 900°C. A significant weight loss took place between 300 and 400°C. The critical temperature for the conversion of resin into carbon was 350°C. Infrared spectroscopy of carbon fibers heat-treated at 350°C showed elimination of methylene bridges and growth of aromatic sheets.

Weight loss and dimensional changes in the bulk glassy carbon were found to be minimal with heat treatments above 1000°C. The electrical resistivity also became minimal at this temperature. Further heat treatment up to 2800°C had no effect, indicating that the carbonization process is probably complete at 1000°C.

The glassy carbon fiber obtained in this way was found to have a very low modulus of 3.5 GN m^{-2} and a tensile strength of 2 GN m^{-2}, which are much lower than the values for PAN carbon fibers. However, the low cost of the source material and the simplicity of the preparation procedure makes them useful materials when mechanical strength alone is not of much importance.

Economy and Lin [49], while investigating the carbonization of Novalac fiber—a phenolic resin fiber marketed by Carborundum Company—observed that the glassy carbon fibers can be oriented to form high-modulus carbon fibers by high-temperature stretching in the temperature range 2200-2750°C.

REFERENCES

1. Bacon, R., and Smith, W. H., *Proc. 2nd Conf. on Ind. Carbon and Graphite*, Soc. Chem. Ind., London, 1965, p. 203.
2. Shindo, A., Report Osaka Industrial Research Institute, No. 317 (1961).
3. Watt, W., Phillips, L. N., and Johnson, W. Engineer (London) *221*, 815 (1966).
4. Kureha Chemical Industry Co., Brit. Patent 1,293,900 (October 15, 1972).
5. Fr. Patent 1,458,725 (November 10, 1966). Nippon Kayaku Co. Ltd.
6. Singer, L. S. Carbon *16*, 409 (1970).
7. Barr, J. B., Chwastiak, S., Didchenko, R., Lewis, I. C., Lewis, R. T., and Singer, L. S. Appl. Polymer Symposia No. *29*, 161 (1976).

7a. Singer, L. S. Proceedings Intern. Symp. on Carbon, Toyohashi, Japan, 1982, 4A01.

8. O'Neil, D. J. Intern. J. Polymeric Mater. *7*, 203 (1979).
9. Ezekiel, H. M. J. Appl. Polymer Sci. Pt. C. *21*, 153 (1973).

10. Ross, J. H. Appl. Polymer Symposia No. *29*, 151 (1976).
11. Shindo, A., Nakanishi, Y., and Sema, I. Appl. Polymer Symposia No. *9*, 271 (1969).
12. Duffy, J. V. J. Appl. Polymer Sci. *15*, 715 (1971).
13. Gutzeit, C. L., U. S. Patent 3,479,150 (1969).
14. Bacon, R., and Schalamon, W. A. *Eighth Bienn. Conf. on Carbon*, Buffalo, N. Y., 1967, summary of papers MI-58.
15. Otani, S. Carbon *3*, 31 (1965).
16. Otani, S. Carbon *5*, 219 (1967).
17. Economy, J. and Wohrer, L. *Proc. 2nd Shirley Intern. Seminar*, Shirley Institute, Manchester, England, Sept. 29–Oct. 1, 1970.
18. Lin, R. Y., and Economy, J. Reprints Am. Chem. Soc. Div. of Organic Coatings and Plastic Chemistry *31*, 403 (1971).
19. Kawamura, K., and Jenkins, G. M. J. Mater. Sci. *7*, 1099 (1972).
20. Br. Patent 1,228,910 (April 21, 1971).
21. U. S. Patent 3,856,593 (October 24, 1974).
22. U. S. Patent 3,903,220 (September 2, 1975).
23. Br. Patent 1,406,378 (September 17, 1975).
24. German Offenleg 2,315,144 (October 11, 1973).
25. Br. Patent 1,274,963 (May 17, 1973).
26. Ezekiel H. M. Appl. Polymer Sumposia No. *9*, 315 (1969).
27. Ezekiel, H. M. Reprints Am. Chem. Soc. Div. of Organic Coatings and Plastic Chemistry *31*(1), 415 (1971).
28. U. S. Patent 3,549,307 (December 22, 1970). Monsanto Co.
29. Celanese Co. U. S. Patent 3,547,584 (December 15, 1970).
30. Economy, J. and Lin, R. Y. Appl. Polymer Symposia No. *29*, 199 (1976).
31. General Electric Co. U.S. Patent 3,928,516 (December 23, 1975).
32. U.S. Patent 3,709,863 (May 7, 1974).
33. General Electric Co. U.S. Patent 3,709,863 (January 9, 1973).
34. General Electric Co. German Offenleg, 2,207,354 (September 7, 1972).
35. Ezekiel, H. M., and Spain, R. G. J. Polymer Sci. Pt. C. No. *19*, 249 (1967).
36. Bourdeau, R. G., and Galasso, F. S. Am. Cer. Soc. Bull. *55*, 785 (1976).
37. Augl, J. M. AD Report No. 780430/5 GA, 1974, p. 37.
38. U.S. Patent 3,635,675 (January 18, 1972).
39. ICI Ltd. Br. Patent 1,283,714 (August 2, 1972).
40. Boucher, E. A., Cooper, R. N., and Everett, D. H. Carbon *8*, 597 (1970).
41. Walker Chemie Gmbh, U.S. Patent 3,732,610 (March 27, 1973).
42. Br. Patent 1,293,900 (October 25, 1972).
43. German Offenleg 1,246,510 (August 3, 1967).
44. Jap. Patent 76 38,525.

45. Jorro, M. A. A., and Ladner, W. R. *Proc. SCI 4th Intern. Carbon and Graphite Conf., London*, Pennard Press, Swansea (1974).
46. Jorro, M. A. A., Ladner, W. R., and Rantell, T. D., Carbon *14*, 219 (1976).
47. Moreton, R., Watt, W., and Johnson, W. Nature 213, 690 (1967).
48. Johnson, W., Phillips, L. N., and Watt, W. Br. Patent 1,110,791 (1968).
49. Economy, J., and Lin, R. Y. J. Mater. Sci. *6*, 1151 (1971).
50. Kawamura, K., and Jenkins, G. M. J. Mater Sci. *5*, 262 (1970).
51. Barnett, I. U.S. Patent 2,913,802 (1959).
52. Boseci Jap. Patent 39/2316 (1961).
53. Johnson, J. W., and Watt, W. Polymer reprints, Am. Chem. Soc. Atlantic City Meeting, 1968.
54. Johnson, J. W., and Watt, W. Appl. Polymer Symposia *9*, 215 (1969).
55. Muller, D. J., Fitzer, E., and Fiedler, A. K. Carbon fibre: Their composites and applications, *Proc. of the Intern. Carbon Fibre Conference, London*, 1971, p. 10.
56. Watt, W., and Johnson, W. *Proc. 3rd Conf. on Ind. Carbon and Graphite*, Soc. Chem. Ind., London, 1971, p. 417.
57. Grassie, N., and McGuchan, R. European Polymer J. *7*, 1503 (1971).
58. Grassie, N., and McGuchan, R., European Polymer J. *8*, 257 (1972).
59. Grassie, N., and McGuchan, R., European Polymer J. *8*, 243 (1972).
60. Grassie, N., and Hay, J. N. J. Polymer Sci. *56*, 189 (1962).
61. Fitzer, E., Heine, M., and Jacobsen, G. Intern. Symp. on Carbon, Toyohashi, Japan, 1982.
62. Fitzer, E., and Müller, D. J. Carbon *13*, 63 (1975).
63. Dorey, Graham Phys. Technol. *11*, 56 (1980).
64. Raskovic, V., and Marinkovic, S. Carbon *13*, 535 (1975).
65. Donnet, J. B., and Ehrburger, P. Carbon *15*, 143 (1977).
66. Raskovic, V., and Marinkovic, S. Carbon *16*, 351 (1978).
67. Raskovic, V., Dezarov, I., and Marinkovic, S. *"Carbon 76" 2nd Intern. Conf. on Carbon*, Baden Baden, 1976, preprints, p. 121.
68. Hantz, R. C. Text. Rex. J. *20*, 786 (1950).
69. Billmeyer, F. W. *Text Book of Polymer Sciences*, 2nd Ed., Wiley-Interscience, New York (1971).
70. Stille, J. K. *Introduction to Polymer Chemistry*, Wiley, New York (1971) p. 172.

71. Fiedler, A. K., Fitzer, E., and Müller, D. J. Am. Chem. Soc. preprints, Div. of Organic Coatings and Plastic Chemistry, 161st National Meeting, Los Angeles, Calif. *31*, No. 1, April 3–5, 1973, p. 142.
72. Standage, A. E., and Matkowsky, R. D. European Polymer J. *7*, 775 (1971).
73. Watt, W. *Proc. 3rd Conf. on Ind. Carbon and Graphite*, Soc. Chem. Ind. London, 1971, p. 431.
74. Bahl, O. P., and Manocha, L. M. Carbon *12*, 17 (1974).
75. Kirby, J. R., Brandrup, J., and Peebles, L. H., Jr. Macromolecules *1*, 53 (1968).
76. Friedlander, H. N., Peebles, L. H., Jr., Brandrup, J., and Kirby, J. R. Macromolecules *1*, 79 (1968).
77. Gupta, S. K. Chem. Age India, *27*(11), 961 (1976).
78. Strauss, S., and Madorsky, S. L. J. Res. Natl. Bur. Stand. *61*, 77 (1958).
79. Monahan, A. J. Polymer Sci. *A-14*, 239 (1966).
80. Burlant, W., and Parsons, J. J. Polymer Sci. *22*, 249 (1956).
81. Myamichi, K. et al. J. Soc. Fibre Sci. Tech. Japan *22*, 538 (1966).
82. Bromley, J. Carbon fibre: Their composites and applications, *Proc. Intern. Carbon Fibre Conf., London*, 1971, p. 3.
83. Hay, J. N. J. Polymer Sci. *A-18*, 2127 (1968).
84. Turner, W. J. Appl. Polymer Sci. *9*, 1245 (1968).
85. Dean, M., Aere Harwell, Private Communication quoted in Watt and Green, paper no. 4, p. 23. Carbon fibre: Their composites and applications, 1971.
86. Bailey, J. E., and Clarke, A. J. Nature 243, 146 (1973).
87. Bailey, J. E., nad Clarke, A. J. Nature *234*, 529 (1971).
88. Goodhew, P. J., Clarke, A. J., and Bailey, J. E. Mater Sci. Engg. *17*, 3 (1975).
89. Coleman, M. M., and Petcavich, R. J. J. Polymer Sci. Polymer Phys. Ed. *16*, 821 (1978).
90. Petcavich, R. J., Painter, P. C., and Coleman, M. M. J. Polymer Sci. Polymer Lett. Ed. *17*, 165 (1979).
91. Coleman, M. M., and Sivy, G. T. Carbon *19*, 123 (1981).
92. Sivy, G. T., and Coleman, M. M. Carbon *19*, 127 (1981).
93. Coleman, M. M., and Sivy, G. T. Carbon *19*, 133 (1981).
94. Sivy, G. T., and Coleman, M. M. Carbon *19*, 137 (1981).
95. Kasatochkin, V. I., and Kargin, V. A. Doklady Phys. Chem. *191*, 303 (1970).
96. Watt, W. Nature *236*, 10 (1972).
97. Watt, W. *Proc. 3rd Conf. on Ind. Carbon and Graphite*, Soc. Chem. Ind., London, 1971, pp. 431.
98. Shindo, A. *Carbon Fibre: Their Composites and Applications*, The Plastic Institute, London, 1971, p. 18.
99. Moreton, R., and Watt, W. Carbon *12*, 543 (1974).

100. Bacon, R., and Schalamon, W. A. Eighth Bienn. Conf. on Carbon, Buffalo, N.Y., 1967.
101. Johnson, W. *Proc. 3rd Conf. on Ind. Carbon and Graphite*, Soc. Chem. Ind., London, 1971, p. 447.
102. Moutaud, G. M., and Duflos, J. L. U.S. Patent 3,322,489 (1967).
103. Yoneshiga, I., and Teranishi, H., Nippon Carbon Co Jap. Patent 2775/70 (1970).
104. Shindo, A. Polymer Reprints *9*, 1333 (1968).
105. Bahl, O. P., and Manocha, L. M. Fibre Sci. Tech. *9*, 77 (1976).
106. Schwenker, R. F., nad Paesu, E. Ind. Engg. Chem. *50*, 91 (1958).
107. Schuyten, H. A., Weaver, J. W., and Reid, J. D. Ind. Engg. Chem. *47*, 1433 (1955).
108. Tamaru, K. J. Chem. Soc. Japan *69*, 21 (1948).
109. Tamaru, K. Bull. Chem. Soc. Japan *24*, 164 (1951).
110. Parks, W. G., Esteve, R. N., Jr., Gallis, M. H., Guescia, R., and Petrarce, A. Mechanism of pyrolytic decomposition of cellulose, Div. of Cell. Chem. Abstracts, 127th ACS Meeting, Cincinnati, Ohio, 1955.
111. Madorsky, S. L., Hart, V. E., and Strauss, S. J. Res. Natl. Bur. Stand. *56*, 343 (1956).
112. Schwenker, R. F., nad Paesu, E. Ind. Engg. Chem., Chem. Engg. Data Series *2*, 83 (1958).
113. Bacon, R., and Tang, M. M. Carbon *2*, 211 (1964).
114. Kanamaju, K., and Koyano, K. J. Appl. Polymer Sci. *3*, 143 (1969).
115. Strong, S. L. Am. Chem. Soc. Div. of Organic Coatings and Plastics Chemistry, preprints *31*, 426 (1971).
116. Bacon, R., Schalaman, W. A., and Strong, S. L. Unpublished technical report (1959). In Chemistry and Physics of Carbon, vol. 9, (Edited by P. L. Walker and P. Thrower) Marcel Dekker, New York, 1973, p. 12.
117. Bacon, R., and Schalaman, W. A. In Chemistry and Physics of Carbon, vol. 9, Unpublished technical report (1968) Ed. P. L. Walker and P. Thrower, Marcel Dekker, New York, 1973, p. 12.
118. Mayer, R. O., Ecker, D. R., and Spry, W. J. U.S. Patent 3,333,926 (1967).
119. Shindo, A. U.S. Patent 3,529,934 (1970).
120. Back, E. L. Pulp. Paper Mag. Can. Technical Section (April 1967) p. 1.
121. Kilzer, F. J., and Broido, A. Pyrodynamics 2, 151 (1965).
122. Ross, S. E. Text. Res. J. *38*, 906 (1968).
123. Peters, E. M. U.S. Patent 3,235,353 (1966).

124. Bacon, R., Cranch, G. E., Moyer, R. O., Jr., and Watts, W. H. U.S. Patent 3,305,315 (1967).
125. Moore, D. R., Ross, S. E., and Tesoro, G. C. U.S. Patent 3,527,564 (1970).
126. Little, R. W. (Ed.) *Flame Proofing Textile Fabrics*, Reinhold, New York, 1947, p. 167.
127. Tang, W. K., and Neill, W. K. J. Polymer Sci. *C6*, 65 (1964).
128. Reeves, W. A., Derkins, R. M., Piccolo, B., and Drake, G. L., Jr., Text. Res. J. *40*, 223 (1970).
129. Morozowski, S. Physical Review *85*, 609 (1952).
130. Bacon, R., and Tang, M. M. Carbon *2*, 221 (1964).
131. Ruland, W. J. Appl. Phys. *38*, 3585 (1967).
132. Bacon, R., *chemistry and Physics of Carbon*, vol. 9, (Edited by P. L. Walker, Jr. and P. Thrower) Marcel Dekker, New York, 1973, p. 1.
133. Ger. Patent 2,027,384 (1970).
134. Jap. Patent 6902510.
135. Ger. Patent 2,161,532 (1972).
136. Jap. Patent 72,10,252.
137. Otani, S. Carbon *3*, 213 (1965).
138. Otani, S. Bull. Chem. Soc. Japan *42*, 1417 (1969).
139. Fr. Patent 2,067,619 (1972).
140. Ger. Patent 2,153,567 (1972).
141. Ger. Patent 2,108,079 (1971).
142. Ger. Patent 2,038,949 (1971).
143. Ger. Patent 2,144,403/4 (1972).
144. Araki, T., and Gomi, S. Appl. Polymer Symposia No. 9, 331 (1969).
145. Otani, S., Yokoyama, A., and Nukui, A. Appl. Polymer Symposia No. 9, 325 (1969).
146. Baker, C., Bentall, R. H., Hawthorne, H., and Linger, K. R. *Proc. 3rd. Conf. on Ind. Carbon and Graphite*, Soc. Chem. Ind. London, 1971, p. 508.
147. Hawthorne, H. M., Baker, C., Bentall, R. H., and Linger, K. R. Nature *227*, 946 (1970).
148. Hawthorne, H. M. *Carbon Fibres: Their Composites and Applications* The Plastic Institute, London, 1971, p. 81.
149. Br. Patent 1,310,769 (March 21, 1973).
150. Singer, L. S. Belgian Patent 797.543 (1973).
151. Fujimaki, H., Kodama, F., Okuda, K., Sakaguchi, Y., and Otani, S. Tanso No. *80*, 3 (1975).
152. Singer, L. S. Carbon *16*, 409 (1978).
153. U.S. Patent 3,976,729 (1976).
154. U.S. Patent 4,026,788 (1977).
155. Volk, H. F. *4th London Intern. Carbon and Graphite Conf.*, Soc. Chem. Ind., London, 1976, p. 183.

156. Tomizuka, I., Isoda, Y., and Amamiya, Y. Tanso No. *106*, 93 (1981).
157. Fitzer, E., Joseph, N., Müller, Th., and Radakrishanan, G. *Carbon 82, The Sixth London Intern. Carbon and Graphite Conf.*, Soc. Chem. Ind., London, 1982, p. 249.
158. Fitzer, E., and Müller, Th. *Carbon 82, The Sixth London Intern. Carbon and Graphite Conf.*, Soc. Chem. Ind., London, 1982, p. 253.
159. Fitzer, E., and Müller, D. J. Chem. Ztg, Chem. Appel, *96*, 20 (1972).
160. Fitzer, E., and Heym, M. Chem. Ind., *21*, 663 (1976).
161. Krzyzanowski, J. Przeglad Wlokiennieczy, p. 112 (1978).
162. Ball, J. R. Chem. Engineer, p. 333 (1977).
163. Green, A. K., and Phillips, L. N. Eng. Applications *1* 59 (1978).

2

Structure of Carbon Fibers

Ever since the introduction of carbon fibers as potential high-strength and lightweight materials in the late 1960s, considerable research efforts have been directed toward understanding their structure, studies have been carried out using several different techniques which could provide information about their micro- as well as macrostructural features. As the microstructure of a carbon fiber is an important factor which influences its strength and fracture behavior, these structural studies have been mainly concerned with understanding the structural differences between fibers and relating them to the observed differences in their physical and mechanical properties.

Several aspects of carbon fiber structure have been examined during the last two decades and considerable agreement has been reached regarding several microstructural features. There is general agreement that carbon fibers are composed of two-dimensional turbostratic graphite crystallites, oriented preferentially parallel to the fiber axis. There is also agreement that carbon fibers have a two-phase structure and have several kinds of micro- as well as macrostructural heterogeneities. Another remarkable feature of these structural studies is the observation that carbon fibers prepared from different precursor materials have many common structural features.

Carbon fibers have been classified into three main types depending upon their fiber structure and the degree of crystallite orientation. Type I carbon fibers are highly graphitized and are characterized by a high modulus (HM). When incorporated into structures, they give the highest stiffness per unit weight. Type II carbon fibers, which are heat-treated to a lower temperature, have a low modulus but a very high strength (HS). Type III carbon fibers have random orientation of the crystallites and display neither the high modulus of type I nor the high strength of type II carbon fibers. Their main advantage is their low cost.

The structural characteristics differentiating these three carbon fiber types are given in Table 1, and their fiber textures are shown in Figure 1 [1].

Since type I carbon fibers are almost pure carbon and generally have reproducible characteristics, most of the structural studies have

Table 1 Carbon Fiber Classification

Carbon fiber type	Heat treatment temperature (°C)	Crystallite orientation	Long distance order
Type I High modulus	> 2000	Mainly parallel to the fiber axis	High
Type II High strength	≅ 1500	Mainly parallel to the fiber axis	Low
Type III Isotropic	< 1000°C	Random	Very low

been carried out on them. Therefore, unless otherwise stated, the reader may consider that type I carbon fibers have been used.

Broadly speaking, the present chapter has been divided into three sections. The first section briefly describes the various techniques which have been and are being used for structural investigations of carbon fibers. Since several review articles and books are available on the technical details of the various techniques, the purpose of this section is to delineate the procedures used for calculation and determination of structural parameters. The results obtained concerning the structural parameters such as crystallite size, preferred orientation, and internal porosity are discussed in the second section. The third section presents a detailed description and discussion of the proposed structural models.

2.1 TECHNIQUES FOR STRUCTURAL STUDIES

The structure of carbon fibers has been widely studied using several different techniques, such as wide and small-angle x-ray diffraction, electron diffraction, transmission, scanning and surface replica electron microscopy, as well as optical microscopy. These techniques have provided useful information about the various micro as well as macro aspects of the fiber structure so that many of the structural aspects are now well understood, accepted, and agreed upon. In addition, these techniques provide qualitative and quantitative information about several structural parameters such as crystallite size and orientation, stack height and width, the shape and size of the pores, and the presence of three-dimensional order.

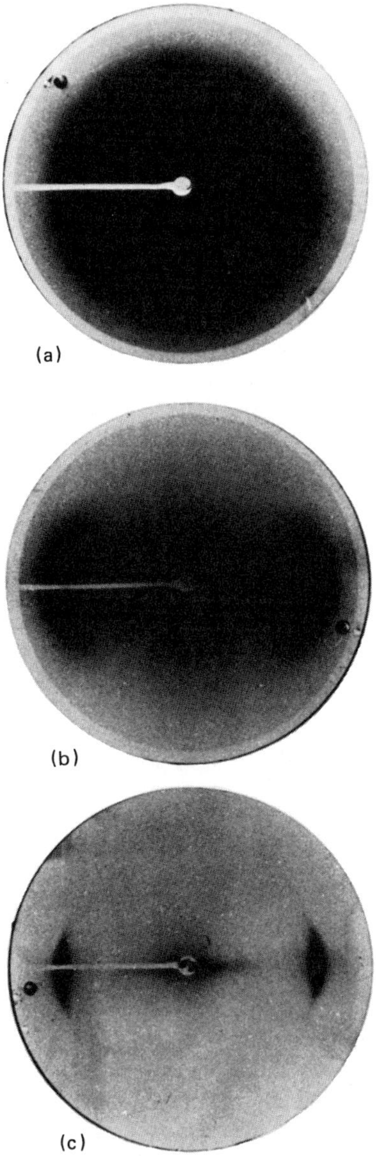

Figure 1 Diffraction pattern of different types of PAN-based carbon fibers. (a) Sigrafil P,E = 40 kN/mm^2 (type III); (b) Sigrafil HF,E = 220 kN/mm^2 (type II); (c) Sigrafil HM,E = 450 kN/mm^2 (type I). (From Ref. 1.)

As the details of these techniques as applied to carbon structure have been discussed in several books [2] and review articles [3–8], only a brief discussion of the more important techniques will be given.

2.1.1 Electron Diffraction

Electron diffraction is more appropriate for studying the surface crystal structure of a carbon fiber than x-ray diffraction since x rays penetrate the entire fiber whereas electrons can only penetrate to a depth of about 100 nm into the surface. Moreover, the electron diffraction patterns are easy to interpret because of their small diffraction angles. Thus, the technique has been widely used for the study of crystallite orientation and the three-dimensional order within carbon fibers [35], and to determine the mean stacking height L_c of basal planes in the core regions of HM carbon fibers [36].

The technique has also been used for studying the orientation and the degree of graphitization as a function of the heat-treatment temperature [35]. The basic theory of electron diffraction is similar to x-ray diffraction and the reader is referred to several good reviews [7,8]. A typical electron diffraction pattern from a fragment of a highly oriented rayon-based carbon fiber is shown in Figure 2. The incident electron beam is normal to the fiber axis and the fiber structure is cylindrically symmetrical. The distribution of scattering intensity may, therefore, be expressed as a function of only two spherical coordinates in reciprocal space, S and ϕ, where S is given by the Bragg equation $S = 2 \sin \theta/\lambda$ and ϕ is the coaltitude angle measured with respect to the fiber axis.

Carbon fibers are mostly nongraphitic and consist of crystallites of turbostratic graphite with random stacking. But this random stacking does not affect the <001> interferences, although it alters the <hKl> interferences resulting in each <hk> row of spots degenerating into a continuous <hk> rod parallel with the C axis. Thus a series of concentric <hk> cylinders, rather than <hKl> hoofs, are found. The rings are formed by the (001) peaks due to fiber symmetry.

The intensity distribution in reciprocal space is obtained by rotation of the <001> spots and the <hk> cylinders about the fiber axis. In the planar section of the rotated intensity distribution the <hk> cylinders produce the <hk> circular bands, each with a pair of <hk> tangential streaks. The inner radius of the bands is sharp and corresponds to the inner radius of the <hk> cylinder in the center of which the intensity is zero. The bands are roughly circles because, over a wide range of angles, the intersection of the <hk> cylinder in the plane of the figure is nearly a circle and one is looking at the walls of the cylinder nearly edge on. The electron diffraction pattern shown in Figure 2 displays <002> and diffuse <100> reflexions.

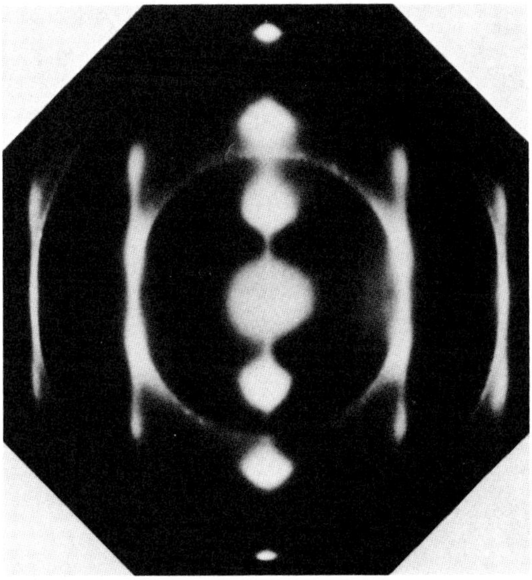

Figure 2 Transmission electron diffraction pattern of a highly oriented rayon-based carbon fiber. (From Ref. 27.)

reflection to be concentrated in two areas, which indicates that the graphite basal planes constituting the fibers are aligned, with their C axis roughly perpendicular to the fiber axis [9–14]. The degree of preferred orientation is generally obtained from the intensity of reflection along the (002) diffraction.

This intensity is measured as a function of ϕ, the angle between the scattering vector and the fiber axis. The angular half-width of this intensity distribution curve measured at the half-maximum intensity (Fig. 3) yields the Bacon orientation function $I(\phi)$ [3,5], which is a measure of the possible misorientation along the fiber axis.

Several other preferred orientation parameters have been recognized. For example, Ruland [15] observed that the broadening of the (002) reflection in the circumferential direction by crystallite size is also important and, consequently, proposed a preferential orientation parameter denoted by q, which is claimed to give a more accurate picture of c-axis orientation, and is obtained from the measured orientation distribution function $I_{002}(\phi)$, using the expression

$$I_{002}(\phi) = \frac{1 - q^2}{1 + q^2 - 2q\cos(2\phi)}$$

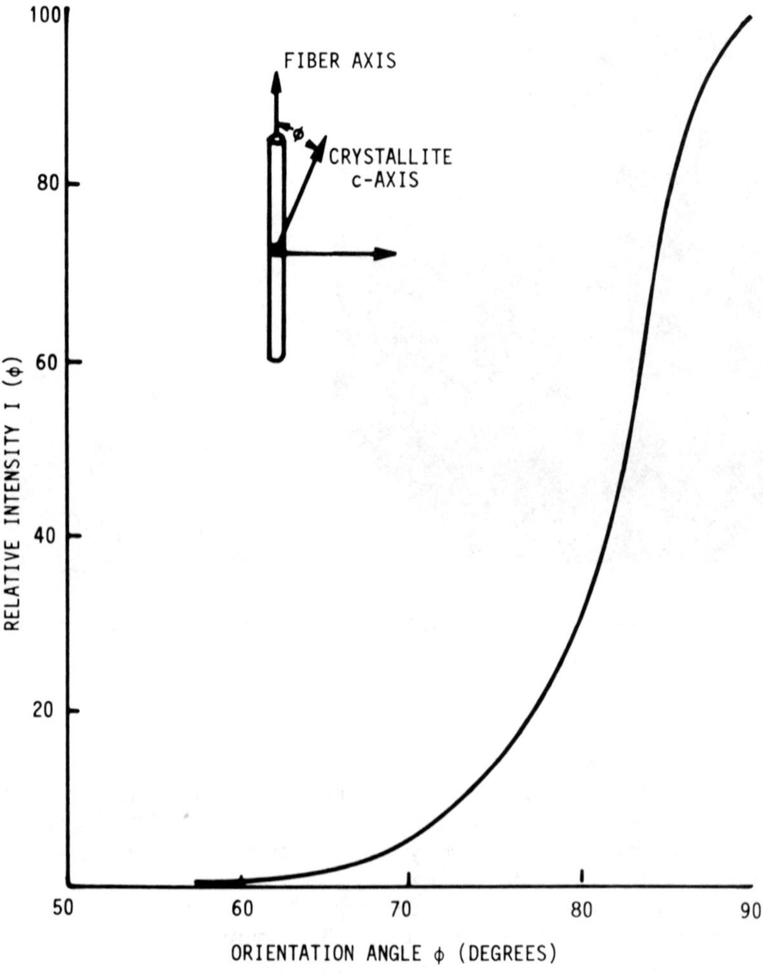

Figure 3 X-ray intensity as a function of orientation angle for Modmor type II carbon fibers. (From Ref. 3.)

2.1.2 X-Ray Diffraction

Wide-Angle X-Ray Diffraction

Wide-angle x-ray diffraction patterns have been used to obtain information about the degree of preferred orientation of graphitic layer planes parallel to the fiber axis, the average stack height L_c, and the average layer with L_a. A typical x-ray diffraction pattern, for a high-strength and high-modulus carbon fiber (Fig. 1) shows (002)

This orientation parameter varies between — 1 for perfect orientation and zero for random orientation. Ruland observed that the above equation approximates well the $I_{002}(\phi)$ function for all rayon-based carbon fibers, except those with extremely high degrees of preferred orientation.

Yamamoto and Yamada [16] defined a preferred orientation coefficient, a_{10}, by the expression

$$a_{10} = \int_0^{\pi/18} \frac{\overline{\pi} \cdot F(\phi)^2 \cdot d\phi}{\overline{\pi} \cdot F(\phi)^2 \cdot d\phi} \quad \frac{\pi}{18} = 10°$$

where

$$F(\phi) = \frac{I(\phi)}{I(0)} = \frac{I_{obs} \cdot (\phi)}{I_{obs} \cdot (0)}$$

and

$$I_{obs}(\phi) = I(\phi) \cdot f^2 \cdot L \cdot A$$

where $I(\phi)$ is the true intensity, $I_{obs}(\phi)$ the observed intensity, f is the structure factor, L the Lorentz factor, and ϕ is the absorption factor. By plotting the angle ϕ against the ratio of intensities $F(\phi)$, the orientation coefficient a_{10} can be calculated. For Thermalon-S graphite fibers the value of this orientation coefficient was found to be 0.935 [16], showing that this carbon fiber has a high degree of preferred orientation.

Another orientation parameter which has been found to be useful for calculating several physical properties such as thermal expansion coefficient [17], magnetic susceptibility [15] and elastic constants [18], can be calculated from the intensity $I_{002}(\phi)$, by using the expression

$$R_{oz} = \frac{I(\phi) \sin^3\phi \, d\phi}{I(\phi) \sin\phi \, d\phi}$$

suggested by Price and Bokros [19] and Gasparoux et al. [20]. Ruland has obtained a relationship between the orientation parameters R_{oz} and q, which holds true in the case of cellulose-based carbon fibers [15].

The crystallite dimension parameters L_c and L_a are generally obtained from the shapes of the principal diffraction lines of a powdered sample. Since the shapes of these diffraction lines depend not only on crystallite size, but also upon structure imperfections, the word apparent is generally used when referring to these dimensions. Thus L_c represents the apparent crystallite stack height and L_a the apparent layer width.

The parameter L_c is obtained from the peak width B at half intensity of the <001> lines, using the Scherrer formula

$$L_c = \frac{k\lambda}{B \cos \theta}$$

where λ is the wavelength of the x-ray, θ is the Bragg's angle, and k is a correction factor of the order of unity. It may, however, be pointed out that broadening of <001> lines can also occur through layer-stacking disorders [21,22] although such effects become predominant only in case of carbon fibers prepared at lower temperature (type II carbon fiber). In the case of type I carbon fibers, the size effects are more predominant. The value of L_c can also be obtained from the full width at half-maximum peak height (FWHM) [23,24]. The variation of L_c with crystallite orientation can be studied by x-ray diffraction pattern of a parallel bundle of carbon fibers, using a diffractometer technique, and measuring the <002> line width as a function of ϕ [5].

The average layer width L_a, the crystallite dimension corresponding to "a" direction is usually determined from the half-peak width $B_{1/2}$ of the |10| or |11| bands using the Warren and Bodenstein [25] formula:

$$L_a = \frac{1.77\ \lambda}{B_{1/2} \cos \theta}$$

Since this formula assumes random orientation of the layers, powdered samples are usually investigated using the cylindrical film camera technique.

Fourdeux et al. [9] have been able to measure both the apparent layer dimensions parallel to the fiber axis ($L_a \|$) and perpendicular to the fiber axis ($L_a \perp$), using a theoretical treatment of the effect of preferred orientation on the (hk) interferences, developed by Ruland and Tompa [26].

Small (Low)-Angle X-Ray Diffraction

The small-angle x-ray diffraction can measure very small density fluctuations which may occur as a result of the existence of pores, voids, or inclusions in the carbon fiber. As the pores in carbon fibers are generally bounded by graphitic layer planes, a sharp boundary is likely to exist between the pores and the dense material, resulting in density variations which may have characteristic dimensions in the range 1–100 nm [5]. Thus, the technique has been used to obtain information about the porous structure of carbons [9,10,27,28]. As the intensity of low-angle scattering is concentrated primarily along the equatorial direction, it is apparent that the pores or voids are aligned along the fiber axis. The dimensions of the low-angle scattering in the equatorial and meridional directions are used to measure the size of the pores [9,29,30,31]. In the case of high-modulus carbon fibers, the pores are generally needle-shaped and are preferentially oriented parallel to the fiber axis. Therefore, small-angle x-ray diffraction can also be used to estimate their length.

Perret and Ruland [32] have identified three other components in the low-angle x-ray scattering of carbon fibers, which also provide useful information about several structural parameters of these materials. One of these components at very small angles is a multiple scattering, perpendicular to the fiber axis, and determines the parameters related to the shape of the filament and the size of its cross-section, as well as the specific surface of the filaments. The scattering at comparatively wider angles is essentially a single scattering comprising two components. One of these components is a scattering of pores, whereas the second component is due to density fluctuations [9,32,33] in the material, which may be caused by fluctuations in the interlayer spacing and statistical variation in the size and shape of layers in a given stack. A third component is a single scattering parallel to the fiber axis, and apparently showed periodic density fluctuations attributed to alternating crystalline and amorphous regions in rayon or partially oxidized PAN fibers [5,34,35,48].

Perret and Ruland [33] calculated the length of the pores from their small-angle scattering data by plotting the product $SB(\phi)$ as a function of S, where $B(\phi)$ is the angular width of the scattering intensity measured at constant S, and S is the magnitude of the scattering vector ($S = 2 \sin \theta/\lambda$). They observed that the function $SB(\phi)$ increased linearly with S. An extrapolation of the linear plot towards zero values of S gave 30 and 100 nm as the minimum lengths of the pores in type I and type II carbon fibers, respectively.

2.1.3 Neutron Scattering

Two crystallographic properties that can be studied by neutron diffraction are the intrinsic width of the (002) Bragg reflection and the variation of the intensity of the reflection with the tilt angle of the fiber. The former is interpreted in terms of the mean crystallite size L_c along the graphite c axis and the latter records the spread of orientation of the crystallite c axis about the fiber axis. This technique has the advantage over the x-ray diffraction technique in that the fiber can be studied in bulk (a complete tow of 10,000 filaments can be used), so that a simple and rapid method of mounting can be used [37].

Saunderson and Windsor [37] studied samples of PAN carbon fibers, heat-treated at five different temperatures from 1000 to 2600°C by

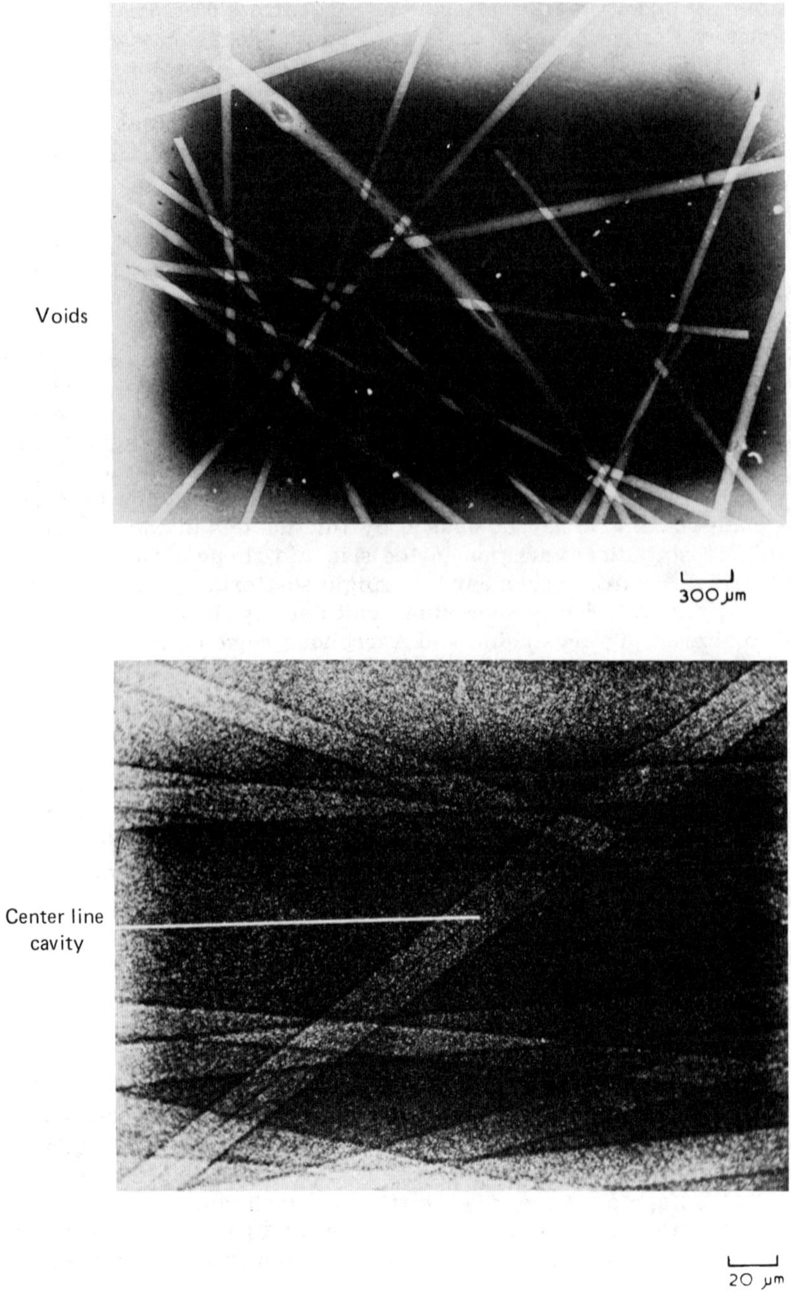

Figure 4 Microradiographs of coal-based carbon fibers showing voids and center line cavities. (From Ref. 39.)

neutron diffraction, and observed that both the crystallite size and the c-axis spread had a marked dependence on the heat treatment temperature. The crystallite size increased linearly whereas the c-axis spread changed in a more complex manner.

2.1.4 Microradiography

A microradiographic technique developed by Sharp [38] has been used by Jarro et al. [39] to investigate the internal structure of coal-based carbon fibers. Radiographs of the fiber filaments were taken using an x-ray microscope. The x-rays were produced electromagnetically, focusing a voltage-accelerated stream of electrons onto a target of copper foil. The size of the electron beam is so chosen as to produce an x-ray source of only 1 μm diameter. The fibers under examination were fixed to a piece of sticky tape mounted above the x-ray probe, and the intensity of the x-ray beam passing through the sample was recorded on a photographic plate. This technique is capable of showing internal and external defects. The microradiographs (Fig. 4) clearly show voids and center line cavities, although there is no indication of any inclusions, which were shown by transmission electron micrographs.

2.1.5 Electron Microscopy

Electron microscopy in its various forms, such as transmission electron microscopy (TEM), scanning electron microscopy (SEM), lattice fringe imaging, and surface replica electron microscopy, is a powerful technique which provides direct observation of the microstructure of carbon fibers and has, therefore, been extensively used. The main advantage of electron microscopy, compared to x-ray, is to produce directly the Fourier synthesis of the object through the image. The Abbe image situated in the back focal plane of the objective lens corresponds to the diffraction pattern, that is, to the first Fourier transform of the object. The optical image corresponding to the second Fourier transform (i.e., diffraction pattern of the diffraction pattern) is formed in the Gaussian plane.

Preparation of a thin sample of carbon fiber for transmission electron microscopy is an art in itself. Several techniques have been suggested and used by different workers. Grinding is one of the simplest and most widely used techniques [29,39,40]. The fiber is ground in a mortar with a pestal and then subjected to ultrasonic treatment, producing a number of very thin flakes which are ideal for electron microscopic examination. However, this technique has been criticized on the ground that the grinding procedure is likely to introduce structural changes. Furthermore, the very thin flakes may originate from the surface portion or from a flaw portion of the fiber and, thus are not representative of the bulk of the fiber.

Flame polishing of the fiber by heat treatment at about 800°C in air [41] can also produce some sufficiently thin specimen areas suitable for electron microscopy. The method is mild and not likely to produce any structural changes in the fiber but it may result in deposition of soot from the flame onto the surface, which may complicate the observation. The "graphite platelets" observed by Coyle et al. [14] in their samples have been attributed to this soot formation [41].

Another technique which has been used to prepare longitudinal and transverse sections of the carbon fiber for electron microscopy is to embed the carbon fibers in a suitable resin and then cut them into thin sections using a microtome. This technique allows the preparation of much larger and reproducible samples without any danger of contamination [40].

Transmission electron microscopy has been used by several investigators [9,10,36,42] to show that carbon fibers have a microfibrillar structure with a high degree of crystallite orientation along the fiber axis. Dark-field micrographs using <002> reflections of the diffraction pattern have been used to estimate the crystallite size [43] and to show that the surface crystallites are larger in size than the internal crystallites, indicating that carbon fibers are of a composite structure [44]. High-resolution transmission electron bright-field micrographs also showed that the microfibrils lie roughly parallel to the fiber axis and are separated by elongated pores [42].

Phase contrast electron microscopy, developed by Heidenreich et al. [45], has been used to show that rayon and PAN-based carbon fibers consist of packets of long parallel stacks of layers spaced 0.34 nm apart [29,45,46].

Surface replica electron micrscopy has been used to study the ultrafine microstructure of carbon and graphite fibers [47]. It has been observed that HM graphite fibers consist of honeycomb axially oriented micropores 6–9 nm apart, separated by long ribbonlike layer packets of turbostratic graphite.

Scanning electron microscopy, although having a smaller resolution power, has the advantage of increased depth of focus and contrast and can, therefore, reveal more clearly the various structural and surface features of carbon fibers and their fracture surfaces. It is capable of achieving a good resolution nearing that of surface replica electron microscopy. The sample preparation technique is simple and, consequently, has been used widely to examine fracture surfaces [48–51], the surface and internal flaws and voids in single fibers and their role in fracture [52], and the structure of fiber-resin composites [53]. SEM micrographs have also been used to measure the radius of gyration of sharp-edged pores and to examine fracture surfaces for the presence of metallic inclusions [54].

2.1.6 Optical Microscopy

Optical microscopy of transverse and longitudinal sections of carbon fibers using polarized light and crossed nicols, or by using the sensitive tint technique has also been used for investigating the microstructure of carbon fibers. The carbon fiber sample is set in a resin using the end-on or the longitudinal position and then polished carefully, using well-lubricated, smooth cloth polishing laps heavily impregnated with diamond paste. This avoids fragmentation of the fibers around the edges and produces a flat fiber cross-section.

As the layer planes of graphite are optically anisotropic, the orientation of basal plane aggregates within carbon fibers can be determined using polarized light. When the polarized light is incident on a fiber cross section normal to the c direction, it is reflected as an elliptically polarized beam which, when viewed through crossed nicols, reveals intensity variations which depend upon the large-scale aggregations of layer planes with similar orientations. Since complete extinction is obtained when the layer planes are either perpendicular or parallel to the direction of polarization of the incident beam, a tint plate is used to determine the orientation of the layers in any particular direction.

Several types of spatial configurations such as circumferential, radial, and random alignments of the crystallites have been distinguished. For example, Butler and Diefendorf [44] and Wicks and Coyle [55] observed a duplex type of structure with almost perfectly oriented material close to the surface in the center of the fiber, and a poorly oriented material in the lobes of the dogbone. Knibbs [56] observed three types of fiber structures, which differed in the orientation of the graphitic crystallites within the transverse plane, depending upon the processing conditions of the carbon fiber. Optical micrographs of fracture surfaces of coal-based carbon fibers showed the presence of blow holes, which appeared to be mineral inclusions from the parent coal [39]. Several different types of transverse layer-plane preferred orientations have also been observed during the examination of graphitized polyvinyl fibers by polarized light optical microscopy [57].

2.2 STRUCTURAL PARAMETERS

2.2.1 Preferred Orientation

Carbon fibers have a high degree of preferred orientation along the fiber axis. This structural parameter bestows on them excellent mechanical properties. For example Ruland [58] and Fourdeux et al. [9] measured the preferred orientation of a number of carbon fibers of different origin in terms of the orientation parameter q (Fig. 5) and observed a direct relationship between Young's modulus and the orientation

(Fig. 7). A similar relationship between Young's modulus and preferred orientation has been observed by several workers [43,59]. In fact, the variation in Young's modulus, which can be measured directly, has been used as a measure of the variation of the degree of preferred orientation in carbon fibers. Shindo [60] was the first to observe that increase in the Young's modulus of PAN-based carbon fibers was related to increase in HTT. This was attributed to an increase in the preferred orientation.

The relative degree of preferred orientation in graphite fibers was measured [61] by measuring the intensity of x-ray reflection along (002) arc by using a standard diffractometer. The half-width of the resulting intensity distribution measured at the point of half-maximum intensity gave a parameter which indicated the relative degree of disorientation of graphite layers from positions of perfect alignment with the fiber axis. This orientation half-width was found to be 25° for carbon fibers with a Young's modulus of 35 GN m^{-2} and decreased with increasing modulus, approaching a nearly constant value of 5° for carbon fibers with a Young's modulus of 560 GN m^{-2}.

Bacon and Tang [62] compared the x-ray diffraction patterns of rayon filaments with varying degrees of molecular orientation before and after graphitization at different temperatures and found that a more highly oriented rayon fiber produced a more highly oriented carbon fiber. This indicates that graphite layers have a strong tendency to form with an orientation parallel to the original cellulose chain direction, so that the cause for high orientation of the carbon fiber lies in the high orientation of the precursor fiber. Ruland [15], on the other hand, observed that the preferred orientation present in the starting material was almost completely lost with the decomposition of the cellulosic structure in the 240–280°C temperature range. The very slight orientation which was left remained almost constant up to a heat treatment of 900°C (Fig. 5). Further heating caused an increase in orientation which improved gradually with rising temperature. A comparison of the preferred orientation in the starting material and in the carbonization products indicated that highly oriented fibers as starting material do not necessarily produce highly oriented carbon fibers. However, in their later work with PAN and rayon carbon fibers, these workers did realize that the orientation of the precursor fiber does affect the orientation of the final carbon product [9]. Barr et al. [23] measured another preferred orientation parameter from the full width at half-maximum (FWHM) using the intensity distribution of x-ray diffraction pattern of HM mesophase pitch carbon fibers and obtained a value of ≅ 30° for raw pitch fibers, which varied between 25 and 36° after thermosetting of the fiber, indicating that the fiber retains its original orientation during the thermosetting step.

Preferred orientation in carbon fibers can be improved either by heat treatment at high temperature or by stretching the fiber during

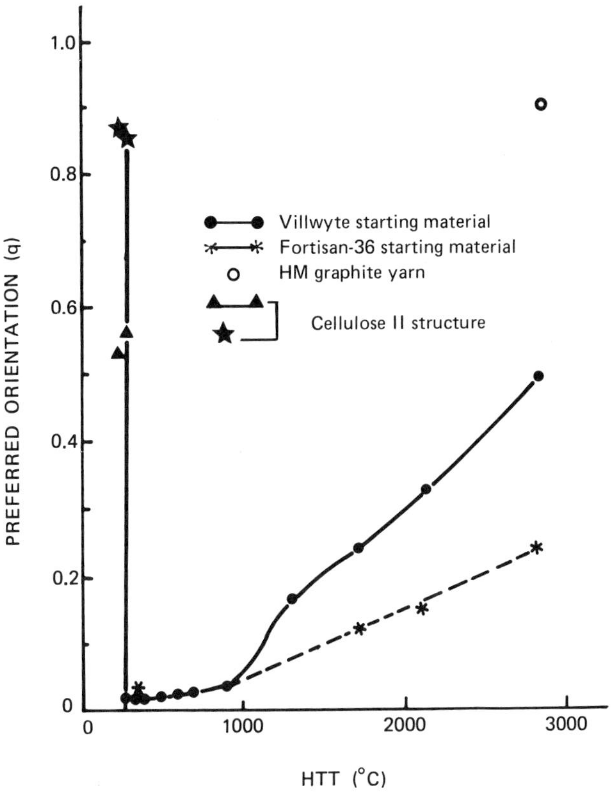

Figure 5 Preferred orientation in carbon fibers as a function of heat treatment temperature. (From Ref. 15, reproduced with permission from Pergamon Press Inc., London; photo courtesy of W. Ruland.)

one of the processing stages. Thus, while in PAN carbon fibers a high degree of preferred orientation is achieved through stretching during the low-temperature preoxidation stage, in rayon fibers it is obtained by high temperature stretch graphitization.

Bright and Singer [24] studied the variation of the orientation parameter FWHM as a function of the heat-treatment temperature (HTT) for two petroleum mesophase pitch fibers; one having a radial transverse structure and the other with predominantly a random transverse structure. The relationship (Fig. 6) indicates significant differences at temperatures up to 1700°C, which become insignificant beyond 1700°C. The fibers with random structure are better oriented initially, but their orientation is as good as those of the radial fibers

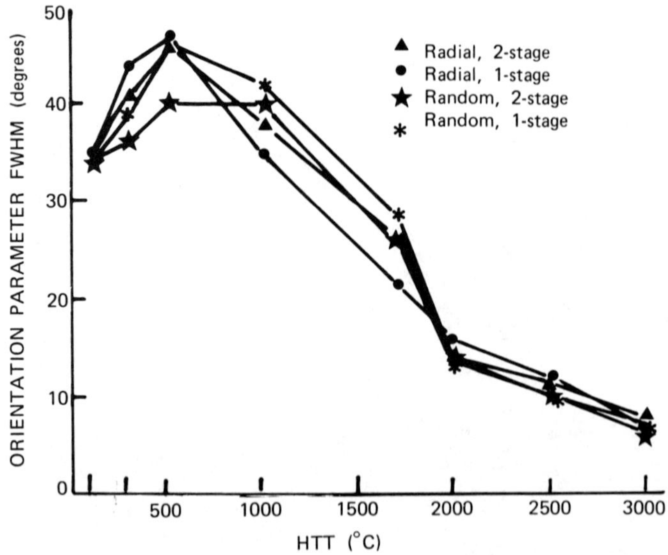

Figure 6 Preferred orientation as a function of heat treatment temperature in pitch-based carbon fibers with radial and random structures. (From Ref. 24, reproduced with permission from Pergamon Press Inc., London.)

at heat treatment temperatures above 2000°C. The FWHM parameter was found to decrease from 30 to 5° on heat treatment to 3000°C [23], indicating an improvement in the degree of preferred orientation. Saunderson and Windsor [63], from their neutron diffraction studies, observed that L_c increased linearly with HTT and was strongly correlated with Young's modulus. Since preferred orientation is directly related to Young's modulus, the data indicate an increase in preferred orientation with increase in heat treatment temperature.

Watt and Johnson [64] observed that the Young's modulus of PAN-based carbon fibers increased with elongation during the oxidation of PAN fibers. This has been attributed to the fact that elongation helps in rendering the molecular chains more parallel to the fiber axis during cross-linking reactions and results in a better orientation of the graphite units. Hawthorne [42] obtained highly oriented high-modulus carbon fibers from isotropic glassy type carbon fibers by subjecting them to axial plastic strain. Figure 7 shows the effect of the degree of strain at 2500°C on the axial preferred orientation of several different carbon fibers obtained from different precursor materials.

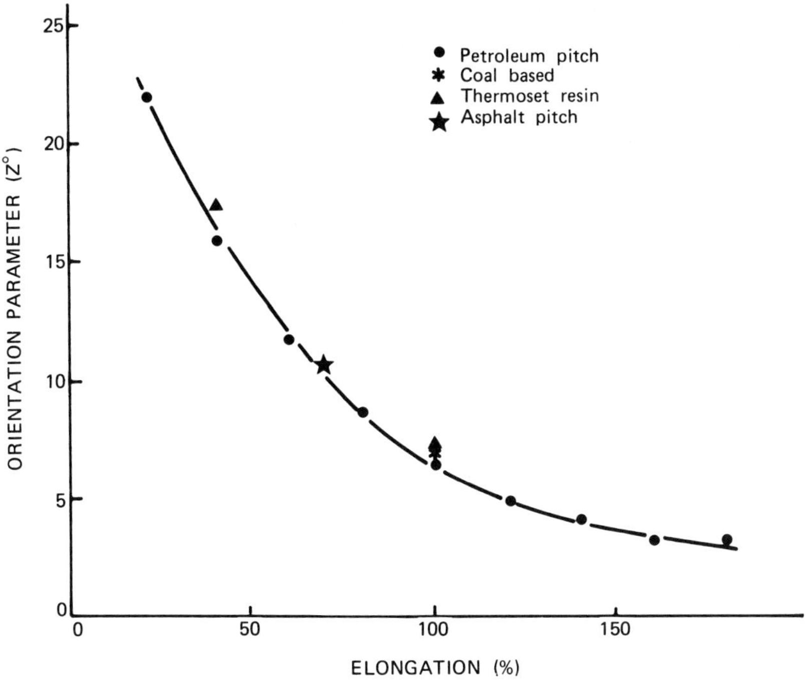

Figure 7 Preferred orientation as a function of strain at the graphitization temperature. (From Ref. 42.)

It is interesting to note that the relationship can be represented by a single curve.

The effect of stretching in the temperature range 2500–2800°C on Young's modulus shows an increase of modulus with extension which in turn indicates an increase in basal plane orientation. Surface replica electron microscopy [47] of carbon fibers produced under high-temperature stress graphitization (Fig. 8) also shows that the carbon fibers are highly oriented.

2.2.2 Crystallite Dimensions

Overwhelming x-ray and electron diffraction evidence suggests that carbon fibers heat-treated to high temperatures (type I) have a two-dimensional turbostratic graphitelike crystallite structure with only a small portion composed of highly oriented crystallite, so that there is

Figure 8 Replicas of external surfaces. (a) Low modulus showing surface ridges; (b) high modulus showing aligned surface ridges. (From Ref. 47, reproduced with permission from Pergamon Press Inc., London.)

no regular structure in the c-axis direction. Thus the crystallite size has generally been studied in terms of two size parameters, the crystallite width, L_a, and the crystallite height, L_c. These parameters depend largely on the heat treatment temperature and on the precursor material.

Johnson and Watt [70] examined the longitudinal sections of a PAN-based carbon fiber heat-treated to 2500°C, and observed that the structure consisted of long narrow units lying parallel to the fiber axis. The crystallites were turbostratic with L_c at least 12 layer planes and L_a in the range 6–12 nm. In the case of the fibers heat-treated to 1000°C, the fibril units were about 80–100 nm across with L_a about 30 nm. Badami et al. [43] calculated the crystallite size to be ≅ 5 nm from line width of (002) reflection, using the Scherrer equation for a PAN carbon fiber heat-treated to 2600°C. The d spacing was found to be 0.339 nm against natural graphite as 0.335 nm. Johnson and Tyson [29] observed the layer spacing to be 0.342 nm for PAN carbon fiber heat-treated to 2650°C just below 0.344 nm commonly assumed for turbostratic carbon.

The variation in crystallite dimensions with heat treatment temperature has been reported by several workers [9,10,24,42,60,63,16].

Table 2 Crystallite Sizes for PAN Carbon Fibers as a Function of Heat Treatment Temperature

HTT (°C)	Crystallite diameter L_a (nm): Eastabrook and Seed [11]	Shindo [60]
1000	3.10	3.19
1500	—	5.06
2500	9.10	11.10
3000	12.15	20.10

The results of Shindo [60] and Eastabrook and Seed [11] show clearly (Table 2) that the crystallite width increases considerably from about 3 to over 12.5 nm as the temperature increases from 1000 to 3000°C. The values of L_a obtained in these two investigations are comparable except at 3000°C where the values are significantly different. Johnson [10], during his investigations of the variation of several structural parameters as a function of HTT by x-ray diffraction for carbon fibers of different origin and prepared under different processing conditions, obtained comparable values of L_a and L_c (Table 3) which

Table 3 Crystallite Dimensions of PAN Carbon Fibers

HTT (°C)	L_c (nm)	L_a (nm)
1000	1.0	2.0
1400	1.8	3.5
2000	3.4	5.4
2400	4.0	6.2
2800	6.0	7.0
2250[a]	20.0	50.0
2500[b]	5.7	4.6
3000[c]	8.6	8.9

[a]Stress graphitization.
[b]Control.
[c]Boron doped.
Source: Ref. 10.

Table 4 Crystallite Width Parallel ($L_a \parallel$) and Perpendicular ($L_a \perp$) to the Fiber Axis

HTT (°C)	Crystallite width (nm) $L_a \parallel$	$L_a \perp$	Preferred orientation parameter $\lvert q \rvert$
	PAN carbon fibers		
1250	3.5	3.4	0.787
2700	11.9	6.4	0.902
	Rayon carbon fibers		
2800	5.9	5.0	0.230
2800[a]	8.1	5.4	0.820
2900	9.1	6.5	0.980
2900	13.0	6.5	0.982

[a]Stretch graphitization.
Source: Ref. 9.

induced him to propose his structural model, consisting of tetragonal crystallites linked end to end by subgrain boundaries. However, both L_a and L_c were found to increase with increased heat treatment temperature. The increase was, however, smaller with heat treatment temperatures below 1400°C. A linear relationship between L_c and HTT was also observed by Saunderson and Windsor [63].

Fourdeux et al. [9] measured the crystallite width L_a parallel ($L_a \parallel$) and perpendicular ($L_a \perp$) to the fiber axis from the width B_{10} of x-ray scattering for several PAN and rayon carbon fibers. The results (Table 4) indicate that the layers are longer in the axial dimension than in their transverse direction. This difference increases with increasing degree of preferred orientation. It can also be seen that HM rayon carbon fibers prepared at the same HTT, but with varying preferred orientation, show almost a constant $L_a \perp$ but an increasing $L_a \parallel$. This difference has been attributed to the straightening of the wrinkled ribbons of graphite layers under the influence of increased thermal versus mechanical treatment, resulting in an increase of the average length. Thus, although L_a is a measure of the length of the straight portions of the layers, it is quite possible that the layers are actually longer than $L_a \parallel$.

The crystallite dimensions are also influenced by the degree of stretch which the carbon fiber received during the various processing stages (Tables 3 and 4). The effect of heat treatment and

Table 5 Structural Parameters for Petroleum Pitch-Based Carbon Fibers

HTT (°C)	d (002) (nm)	L_c (nm)	L_a (nm)
1000	0.372	1.10	3.92
1500	0.358	1.52	7.13
2000	0.347	2.52	7.45
2500	0.346	2.69	6.85
2750	0.344	3.20	7.45
2850	0.341	3.33	6.73
	Strain Graphitized (80%)		
2350	0.344	3.75	—
2650	0.341	4.40	—
2900	0.340	6.20	—

Source: Data from Ref. 42.

Table 6 Structural Parameters for Petroleum Pitch-Based Carbon Fibers Strain-Graphitized to Various Extents

	2850°C[a]			2500°C[b]	
% Elongation	d_{002} (nm)	L_c (nm)	L_a (nm)	d_{002} (nm)	L_c (nm)
0	$.341^5$	3.3	7.2	.345	2.55
20	.342	2.85	7.8	.342	2.75
40	.342	4.4	8.0	.341	3.05
60	$.342^5$	4.75	8.9	.343	3.15
80	.342	5.1	7.4	.342	4.15
100	.342	6.0	10.8	.343	4.3
120	$.340^5$	8.9	13.6	.342	6.3
140	–	–	–	.342	6.3
160	$.341^5$	11.6	–	–	–
180	.341	14.5	–	–	–

In this table, the exponent 5, after d_{002} and L_c means: values from Ref. 112.
[a]Debye-Scherrer patterns from powdered samples.
[b]Debye-Scherrer patterns from whole fibers mounted perpendicular to beam.
Source: Data from Ref. 42.

stretching on the variation of crystallite dimensions in the case of pitch-based glassy carbon fibers (Table 5) shows that, whereas L_a and L_c increase with increasing HTT, the layer spacing is found to decrease. Comparison of the data in Table 5 shows that the introduction of basal-plane preferred orientation into the fiber by stretching allows an easier approach to a more graphitic state at lower temperatures. The effect of elongation during graphitization on the fiber processed at the two temperatures 2500°C and 2850°C is shown in Table 6. The data clearly indicate that high-temperature stretching has a marked influence on the apparent stacking size and there is some evidence that the apparent layer dimensions and perfection increase at very high elongations. However, there is no significant effect on the interlayer spacing at the high temperature. Similar effects of stretching on the microstructural parameters have been reported for PAN-based [64] and rayon-based [66] carbon fibers.

X-ray studies of PAN carbon fibers after boron doping and neutron irradiation [67] showed that the crystallite size increases in the case of the boron-doped samples while it remains more or less unchanged by neutron irradiation. Measurement of the x-ray line broadening of the powdered samples showed that L_c increased from 5.7 to 8.6 nm and L_a increased from 4.6 to 8.9 nm after boron doping of the sample (Table 3). Electron microscopy at medium resolution [68] showed a wide distribution of crystallite size in the case of HM fibers heat-treated to 2800°C.

Bennet et al. [36] compared the mean stacking height of basal planes L_c, in a high-modulus carbon fiber, obtained by high-resolution electron microscopy, and comprehensive electron and x-ray diffraction analyses. Direct measurements using bright-field and dark-field images yielded considerably lower values than those obtained from equatorial section analysis of the (002) profile (Table 7). However, when the decrease in crystallite size with misorientation from the fiber axis was assessed quantitatively, the diffraction values agreed very well with the value obtained directly from high-contrast, high-resolution dark-field images. This has been attributed [36] to the shortcomings of the conventional diffraction method which takes into account the (00l) profile only within the finite angle of azimuthal misorientation $(90 - \phi)$ controlled by slit length. In addition, these workers question the use of powder technique for microstructure studies because it does not result in the random dispersion of all crystallites. These workers suggest that the specimen rotation technique is better and should be preferred to obtain "meaningful" values of L_c by x-ray diffraction.

Bright and Singer [24] observed that for mesophase pitch carbon fibers, the relationship between the crystallite size L_c and the heat treatment temperature [9] showed an apparent decrease in L_c at low HTT and a rapid increase above 1000 to 2500°C. The radial structured fibers attained a higher value of L_c than the random structured

Table 7 Comparison of Crystallite Size Measurements by Different Techniques

Technique	L_c (nm)
High-resolution electron microscopy	
1. Multibeam bright-field image at Gaussian focus	4.5
2. Multibeam bright-field image at maximum fringe-contrast focus	4.5
3. 000–002 symmetric bright-field image	4.7
4. 002–004 symmetric dark-field image	4.0
5. 002 axial dark-field image	3.5
Electron diffraction	
$(90 - \phi) = \pm 2°$	5.2
$(90 - \phi) = \pm 90°$	4.3
Mean defect-free distance	5.9
X-ray diffraction	
$(90 - \phi) = \pm 1°$	5.1
$(90 - \phi) = \pm 90°$ (rotated specimen)	3.8
Powder photograph	5.9
Computation	
Distortion-free stacking size	8.3

Source: Data from Ref. 94.

fibers, although the differences were insignificant below 1700°C. The latter fibers leveled off at about 2500°C showing little, if any, further improvement in crystallite size.

The effect of orientation of the crystallites on the apparent crystallite height L_c, as studied by Perret and Ruland [33], is shown in Figure 9 for PAN and rayon carbon fibers. The orientation of the crystallites is represented by the angle ϕ between the layer normals and the fiber axis. The breadth of the (002) line B_{002} represents the reciprocal of the apparent size L_c according to the Scherrer formula. The L_c values are highest for those layers that are most nearly paral-

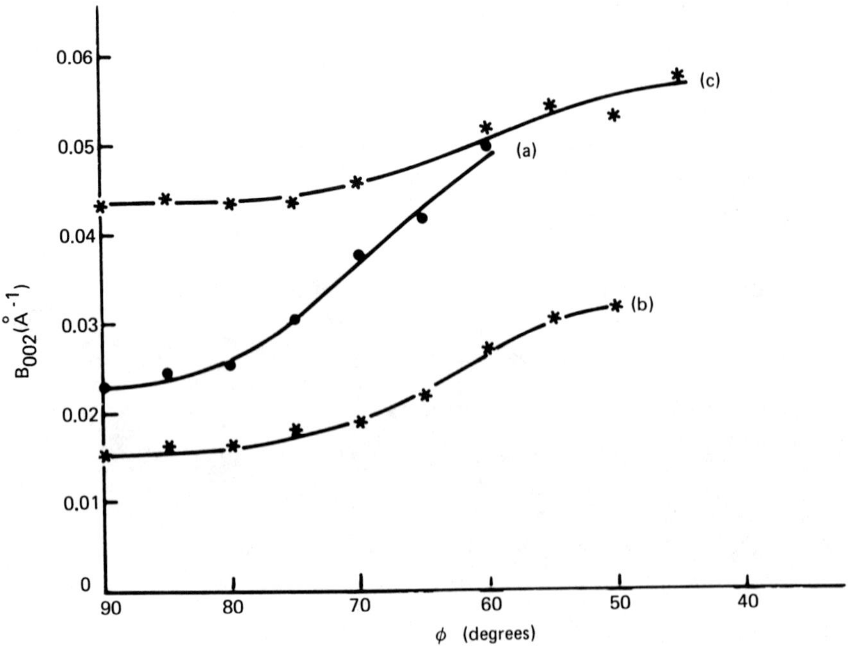

Figure 9 Breadth (B_{002}) as a function of orientation angle (ϕ) for (a) rayon-based Thornel 25, (b) type I PAN-based and, (c) type II PAN-based carbon fibers. (From Ref. 33; photo courtesy of W. Ruland.)

lel to the fiber axis. The effect is more apparent in case of the rayon fiber. This has been attributed to the existence of ribbonlike graphite layers stacked in a parallel arrangement to form microfibrils, so that the thickness of the microfibril is greatest where they are nearly parallel to the fiber axis and is least where they branch off into off-axis direction [5].

The apparent L_c values for carbon fibers heat-treated at lower temperatures are considerably lower than those for the type I carbon fibers [9,63,69]. Since imperfect layer stacking in these carbon fibers may contribute to the broadening of (002) lines, the true layer stack height is always greater than the calculated L_c.

2.2.3 Pores

The density and the crystal density of carbon fibers vary between 1.74 and 1.86 g cm^3 and 2.03–2.21 g/cm^3, respectively, depending upon their Young's modulus [70]. Thus, carbon fibers have an ap-

Table 8 Internal Specific Surface and Pore Width from X-Ray Diffraction Studies for PAN Carbon Fibers

Heat treatment temperature (°C)	l_p (nm)	S_v (Mm^{-1})
600	0.33	—
800	0.48	—
1000	0.88	1450
1250	1.36	960
1400	1.16	1100
1600	1.12	925
1700	1.14	—
1800	1.28	980
1900	1.82	—
2000	2.54	485
2200	2.46	485
2250[a]	2.6	—
2400	2.64	363
2500	—	—
2500	2.8	380
2800	2.8	340
2800	1.	495
2800	2.60	365
2800	3.01	330
3000[b]	4.0	280

[a]Stress-graphitized.
[b]Boron-doped.
Source: Data from Refs. 10 and 29.

preciable porosity (≅ 16–18%). Most of this porosity exists in the form of sharp-edged needle-shaped pores with very small diameters. As the diameters of these pores are small and as their presence creates density transition in the material, estimates of their size and shape are generally obtained from small-angle x-ray scattering.

Ruland and Perret [32,33] and Ruland [71] observed that small-angle x-ray scattering from rayon- and PAN-based carbon fibers contained two components. The predominant component is a scattering of needlelike pores with diameters in the 1–2 nm range and lengths at least 20–30 nm. The pores have a preferred orientation of their needle axes parallel to the fiber axis and the angular distribution of their needle axes very similar to that of graphitic layers.

These workers also studied the influence of heat treatment temperatures and stretching on the size of the pores [9,35] and observed that the average diameter of the pores perpendicular to the fiber axis was a function of the heat treatment temperature and varied from 0.6 nm at 900°C to 2 nm at 2900°C. The variation in pore diameter was almost linear up to about 2000°C, but the diameter increased more rapidly at higher temperatures. The average pore diameter was found to be independent of the starting material and stretching. The porosity remained almost unchanged by increasing heat treatment temperature, as well as by stretching at lower degrees of stress, but decreased to about one-third of its original value for higher degrees of stretch in fibers with Young's modulus in the range 142–460 GN m^{-2}. This decrease in porosity without an appreciable decrease in pore diameter or pore length indicates that the high degree of stretch causes collapse rather than a gradual closure of the micropores. The micropores were in general inaccessible to helium and contributed to about 30 percent of the total volume.

The lobe-shaped low-angle x-ray photograph of PAN carbon fiber obtained by Johnson and Tyson [29] has been attributed to the existence of pores which introduce a marked density fluctuation, causing an intense scattering. The data [29,10,30] revealed that the pores were thin and needlelike, being longer in general than the crystallites which bounded them. The data for internal specific surface (S_V) and the width of the pores (lp) for PAN carbon fibers of different origin, and heat-treated to different temperatures and under stress are given in Table 8. It is evident that the pore width increases while the internal surface decreases with increase in the heat treatment temperature. This shows that, as graphitization proceeds, some of the smaller pores are eliminated as suggested earlier [9,35]. The discontinuity in lp at about 1900°C has been attributed [68] to a change from small distorted crystallites with many small pores to broader, more perfect crystallites separated by larger pores. Electron microscopy also indicated a wide variation of pore sizes in HM fibers heat-treated beyond 2800°C.

Hot stretching caused only a slight increase in the size of the pores (Table 8) but the pores became more uniform [10]. Comparison of Figures 10a, and b indicates that the pores in the case of the stretched fibers are relatively more uniform, longer, and narrower compared with normal fiber heat-treated to 2800°C. High-resolution TEM [42] also suggested that highly strained fibers have longer pores compared to unstretched fibers.

The low-angle parameters [10] for boron-doped carbon fibers (Table 3) indicate a definite improvement in the orientation of the pores. But a slight oscillation observed in the Mering plots indicates that the pores in the boron-doped sample were not as uniform as in the case of hot-stretched fibers. Although the treatment of carbon fiber with boron decreased the number of pores, it increased their average diameter [67].

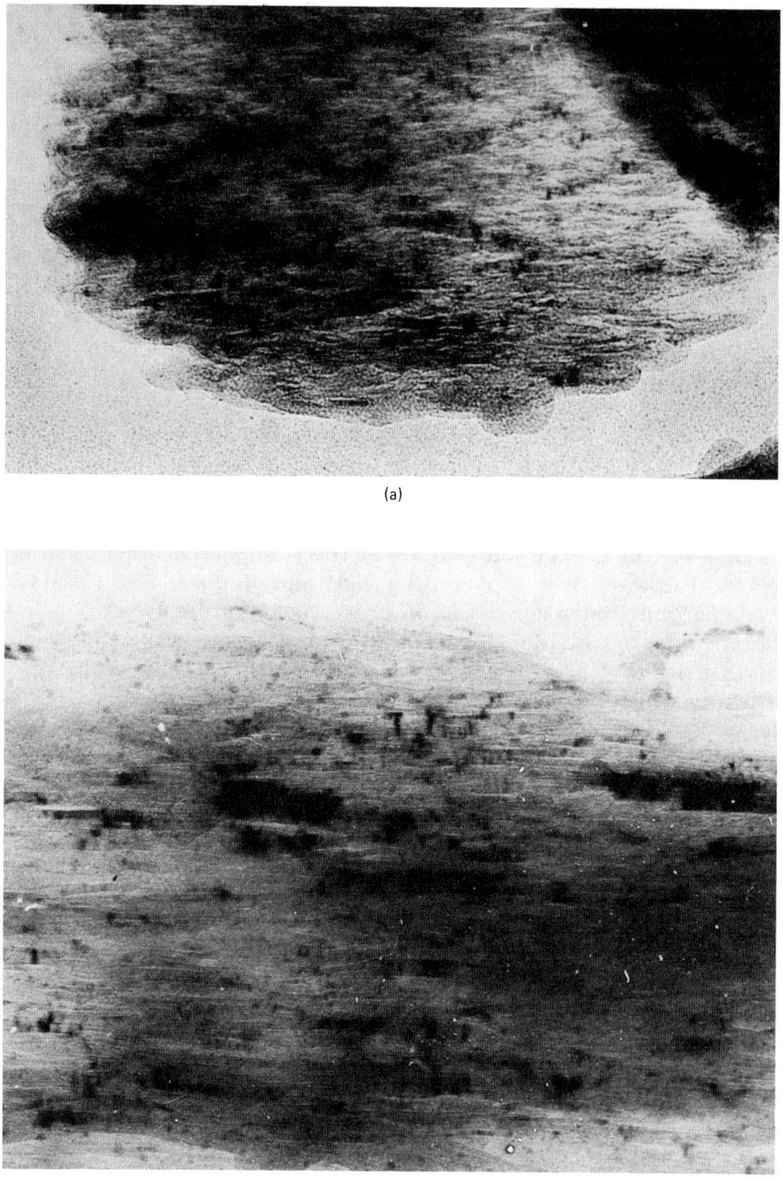

(a)

(b)

Figure 10 Effect of hot-stretching on the size and shape of pores. (a) PAN-based carbon fiber heat-treated at 2800°C (× 270,000); (b) PAN-based carbon fiber stress graphitized (× 200,000). (From Ref. 10, reproduced with permission from the Plastics Institute, London.)

Bacon and Silvaggi [47], from their examination of high-modulus rayon carbon fibers by x-ray diffraction and surface replica electron microscopy, found that these fibers consisted of honeycomb axially oriented micropores 6–9 nm across, separated by long ribbonlike layers of turbostratic graphite. The low-modulus carbon fibers were almost featureless and a well-defined pore structure developed only upon heat treatment of the carbon fiber at 2000°C. Porosimeter measurements showed that graphitization caused a decrease in the number of small pores and developed larger pore sizes. These large pores became closed during their development and were thus inaccessible, even to helium.

On stress graphitization, the fiber structure became more highly oriented and it was possible to identify the individual pores extending several hundred nanometers in length.

2.3 STRUCTURAL MODELS

2.3.1 Microfibrillar Model

X-ray and electron diffraction studies of the fiber structure of PAN-based carbon fibers, first reported by Watt and Johnson [64], showed that these carbon fibers consisted of long primary units lying parallel to the fiber axis and bonded together to form a stretched network of branched fibrils that apparently run the full length of the fiber. The fibrils have a width of about 10 nm. X-ray diffraction studies showed that the crystallites are turbostratic in structure, with L_a varying between 6–12 nm and L_c having at least 12 layer planes. Essentially the same conclusions were reached by Badami et al. [43] and Brydges et al. [31] in their studies of high-modulus PAN carbon fibers by x-ray diffraction and TEM. The size of the crystallites calculated from the line widths of (002) reflection, using the Scherrer equation, was 5 nm. The basal planes were highly oriented along the fiber axis with a spread of about ± 10°. The d spacing was found to be 0.339 nm, approaching more closely the value for natural graphite (0.335 Å). The crystallites were arranged in chains inside fibrils of about 50 nm diameter.

The first detailed proposal of the microfibrillar structure, however, was first provided by a series of studies [12,13,35,72–74] of x-ray diffraction and electron microscopic procedures on rayon-based carbon fibers. According to these workers, the basic unit of the structure consists of ribbon-shaped monatomic layers of sp^2-type carbon, with an average width of 6 nm and a length of several hundred nanometers. The ribbons display irregular contours which may contain holes. There is no correlation between the general direction of the ribbon and the direction of the a axis of the internal two-dimensional hexagonal structure of the ribbons. A certain number of these ribbons run

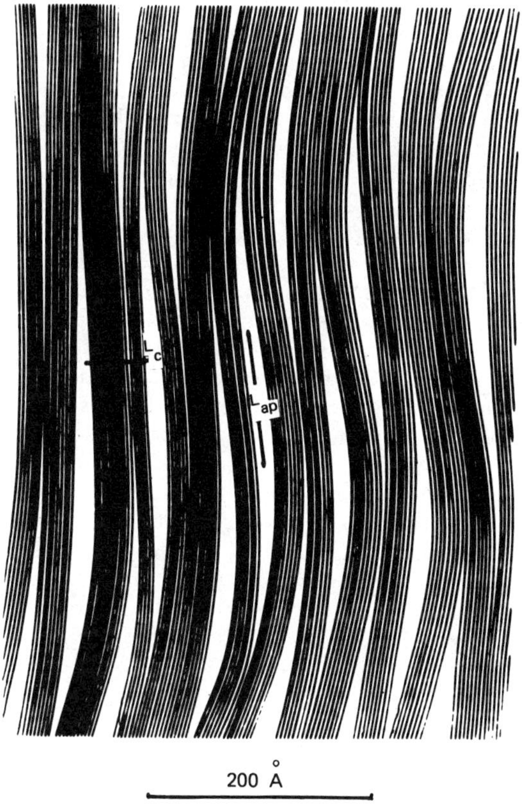

Figure 11 Schematic representation of fibrillar carbon fiber structure. (From Ref. 33; photo courtesy of W. Ruland.)

parallel to form microfibrils which have a preferred orientation parallel to the fiber axis. The ribbons are slowly undulating, generally following the direction of the fiber axis, and contain straight parts with lengths in the 6–13 nm range, depending upon the preferred orientation. The ribbons show a high degree of parallel stacking and pass smoothly from one turbostratic domain of stacking to the other, resulting in a turbostratic disorder. Since the contours of the microfibrils do not match (i.e., they are wrinkled), the space cannot be completely filled and creates a number of needlelike voids. These voids are long (20–30 nm) and thin (av. cross-section 1-2 nm) and their preferred orientation follows closely the orientation of the straight parts of the ribbons [9]. A schematic representation of the structure proposed by Perret and Ruland [33] is given in Figure 11.

Johnson and Tyson [29,30], based on their investigations by x-ray diffraction and electron microscopy, presented a somewhat different model for the structure of graphitized PAN-based carbon fibers, which was further supported by direct lattice resolution of layer planes [40]. The idealized structure suggested by these workers consists of arrays of misoriented turbostratic graphitic crystallites stacked approximately end to end in a columnar arrangement forming tilt and twist boundaries between them. The crystallites are separated by intercrystallite boundaries and sharp-edged voids. The average width of the crystallites was 65 Å (6.5 nm) and the mean width of the voids was estimated to be less than 10 Å (1 nm). In addition to the turbostratic graphite, the carbon fibers also contained small amounts of more perfect three-dimensional graphite, and this component constituted a larger portion in fibers that had been crystallized in the presence of Ni [75,76]. Extinction bands observed in the turbostratic phases of both fiber types were suggested to result from subgrain boundaries with predominantly twist components. The presence of tilt boundaries was also inferred and the boundaries were assumed to comprise nonbasal screw and edge dislocations. A schematic sketch showing tilt and twist boundaries running transverse to the fiber axis is shown in Figure 12.

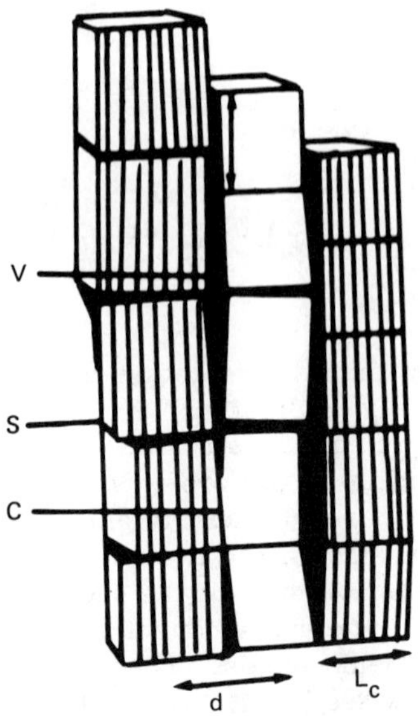

Figure 12 Schematic carbon fiber structure with columnar arrangement of misoriented turbostratic graphite crystallites forming tilt and twist boundaries. (From Ref. 29.)

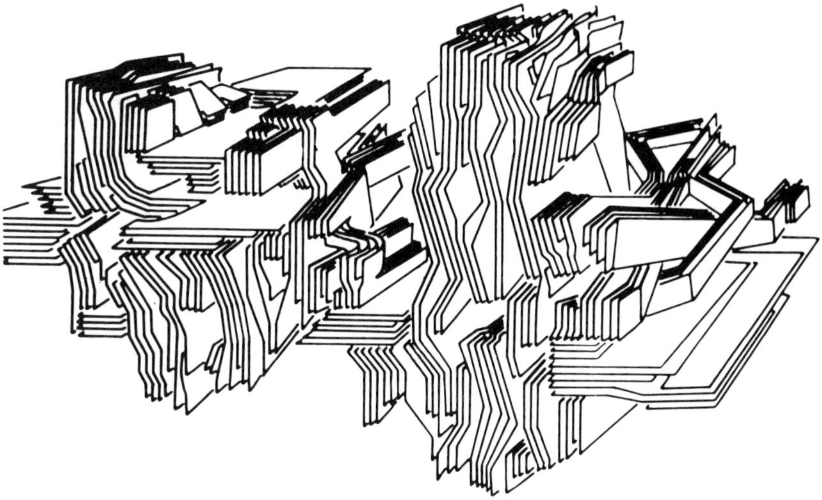

Figure 13 Schematic representation of high-modulus carbon fiber structure. (From Ref. 68; photo courtesy of D. J. Johnson.)

In their later work with PAN-based carbon fibers with exceptionally high modulus, obtained by boron doping at 3000°C, Johnson et al. [68] observed high values of L_c, L_p (Porod's distance of heterogeneity), and a low orientation. The x-ray parameters were exceptional and the resolution of 0.34 nm lattice in the electron microscope showed complex subgrain boundary of the type postulated in the earlier model [29]. On the basis of examination of tilted specimen, together with lattice resolution, these workers suggested another structural model for HM carbon fibers (Fig. 13). This model consists of interlinked crystallites, sharp-edged voids and imperfect stacking with twist boundaries and a vertical arrangement of subgrains. The model depicts the essentially unstable turbostratic form of structure which can recrystallize by twisting and bonding.

In order to resolve the difference in the structure of rayon-based carbon fibers proposed by Fourdeux et al. [9], Ruland [15], and Perret and Ruland [33], and of PAN-based carbon fibers suggested by Johnson and co-workers [68], studies [9,33,15] examined the structure of a number of carbon fibers of different origin (pitch, rayon, PAN) and having different heat treatment temperatures, and observed a marked resemblance in their basic structural features. For example, high-resolution dark-field electron micrographs showed möire patterns both for the |002| and |10| interferences which confirmed the model of slowly undulating ribbons of sp^2 carbon, with an average width of 5–7 nm depending upon HTT, as the basic structural unit. Dark-field micrographs with the |103| interference of partially graphitized carbon fiber indicated that AB ordering was restricted to a small number of layers and distributed at random in the layer stacking. X-ray wide-angle scattering variation both of (00l) and (hk) line widths was interpreted in terms of structural parameters related to the sequence of straight and curved regions in the microfibrils [9]. Thus, there was clear evidence for similarity in the structure of PAN- and rayon-based carbon fibers. The only differences which these workers observed was a considerably lower volume fraction occupied by voids in the case of PAN carbon fibers, which they attributed to differences in the micromorphology at critical stages of the pyrolysis process. According to these workers [9,15,33] the basic structure was established much earlier in the temperature range 1000–1200°C and showed an evolution towards larger domain sizes, higher preferential orientation, and increased in relative perfection with increasing HTT and mechanical treatment, without any changes in the basic structural features. The equivalence in the relationships between cross-section of the voids and the heat treatment temperature, and between Young's modulus and the preferred orientation for different carbon fibers [9], suggested that the carbon fiber structure was a semistable state of the fibrous carbon which can be approached from a variety of starting materials and through different pyrolytic processes, provided that certain conditions which relate to those of growth of ribbons at low temperatures were fulfilled.

These workers [9,33] found no evidence for the existence of sharp grain boundaries transverse to the general directions of the ribbon or for the existence of amorphous carbon, as was suggested by Johnson and Tyson [29,30]: the only boundary regions which they observed were voids separating adjacent packs of ribbons parallel to the layer planes and the regions formed by the atoms located at the lateral boundaries of the ribbons. The structure proposed by Johnson and Tyson was further contradicted by studies on the intercalation of potassium in type I carbon fibers [33,77]. The incorporation of potassium did not affect the preferred orientation of the number of carbon layers per stack, thus excluding the existence of any effective cross-linking perpendicular to the layer planes. This is in contradiction with the assumption of twist boundaries perpendicular to the fiber axis, and supports the "wrinkled ribbon" structure. The absence of any kind of normal or subgrain boundaries was also observed by Hugo et al. [46] in their microstructure studies of Thornel 50 rayon-based carbon fiber by TEM.

The microfibrillar structure was further supported by studies of strain-graphitized pitch-based carbon fibers using replica transmission electron microscopy, scanning electron microscopy, and electron diffraction [42]. High-resolution transmission electron micrographs showed evidence of microfibril structural units lying roughly parallel to the fiber axis and separated by elongated pores, the length of which appeared to be greater in the more highly strained fibers.

Voet et al. [78], while studying the dynamic mechanical properties of graphite fibers in a low frequency range, indicated the presence of relatively large, independently mobile fiber constituents, which appear to correspond to the slowly undulating ribbons recognized as basic carbon fiber structural components. Mencik et al. [79] during their examination of longitudinal microtome sections of two PAN-based commercial carbon fibers, used a different direction with respect to the fiber axis by TEM and selected area electron diffraction (SAD). While agreeing that carbon fibers have a fibrillar structure, these workers suggested caution in interpretation of secondary morphological features with dimensions greater than the fibril range (i.e., above 15 nm), which were attributed to knife-induced artifacts.

The effect of strain rate and heat treatment temperature on microstructural parameters, such as preferred orientation, L_c and L_a, on different types of carbon fibers was studied by several workers [9, 80,42,10]. The results (Table 3, 5, and 6) could be very well explained on the basis of the wrinkled ribbon model. According to Perret and Ruland [80] the influence of strain could be explained by unwrinkling of the ribbon-shaped carbon layers and/or accompanied by gliding of groups of ribbons, depending upon the strain rate. The effect of heat treatment is to reduce the irregularities of the bound-

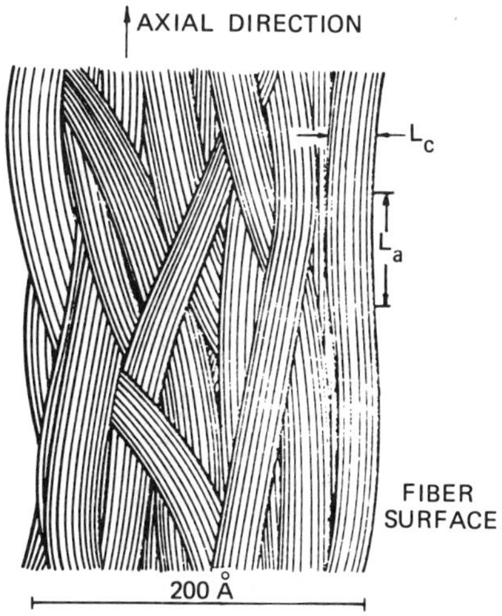

Figure 14 Basketweave ribbon structure model for carbon fiber. (From Ref. 81.)

Figure 15 TEM micrograph showing a sheetlike structure for a high-modulus carbon fiber. (From Ref. 81.)

aries and the chemical interactions which stabilize the wrinkled state so that the ribbons take up the straight configuration. This results in an increase in preferred orientation as well as lattice diameter L_a parallel to the fiber axis. Since the straightening can lead to closure of voids between stacks of ribbons, the stack size L_c could also increase, which is certainly the case when heating and stretching are combined.

Diefendorf and Tokarsky [81] examined HM carbon fibers using TEM and showed that carbon fibers have a wrinkled ribbon structure, the ribbons varying in thickness between 13 and 30 layers and in width between 4 and 9 nm. A schematic diagram of their basketweave ribbon structure is given in Figure 14. The ribbons have almost zero amplitude and are essentially parallel to the fiber axis. These workers suggest that the structure could be better described as a wrinkled sheet. A bright-field TEM micrograph showing the sheetlike structure of a very high-modulus carbon fiber is shown in Figure 15. Such a sheetlike structure has also been shown in several other TEM pictures and in the replica electron microscopy of radio frequency sputtered-etched samples [81].

Although there is no complete agreement, the structure seemed at least generally understood so far as the basic unit and the features in a direction parallel to the fiber axis and perpendicular to the layer planes were concerned. However, considerably less information was available about the arrangement and interconnections of the microfibrils in the transverse direction (i.e., in the section perpendicular to the fiber axis). Fourdeux et al. [9] considered this question and, although they could not set up a definite model for this section, they indicated some structural features which should be taken into account when constructing such a model. They have indicated some evidence for a short-range order in the packing of adjacent stacks of ribbons. There was a tendency for lateral ribbon boundaries to contact the lateral boundaries of an adjacent ribbon rather than the surface of the graphitic layer plane (c plane). This evidence resulted from the high probability of the occurrence of möire fringes in (002) dark-field micrographs [9], as well as from the optical anisotropy of cross-sections of fiber reported by Butler and Diefendorf [44,82]. Fourdeux et al. [72] further observed that the lateral dimensions of the ribbons were nonuniform, indicating the probability that some ribbons run across two or more microfibrils tying them together. Some evidence suggestive of the interweaving of ribbons or small fibrils across the fiber, from one fibril to the next, was also available from the electron microscopic investigations of strained graphitized pitch fibers [42]. These links could account for the relatively high strength of the fibers by providing a measure of transverse bonding across the fiber and resisting interfibril movement under axial stress.

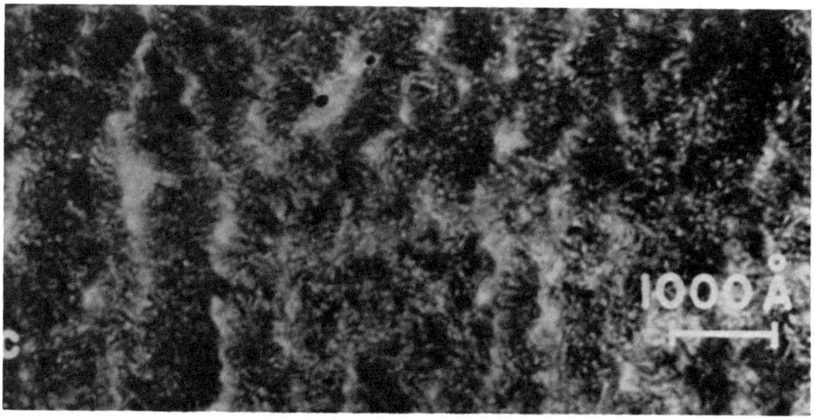

Figure 16 Bright-field electron micrograph of a transverse section of a rayon-based carbon fiber. (From Ref. 5.)

Electron micrographs of transverse sections of rayon-based Thornel 50 fiber [5,47] reveal a multiplicity of white dots on a dark background (Fig. 16), which has been interpreted by Bacon and Silvaggi [47] as being due to honeycomb axially oriented micropores 6–9 nm apart, separated by packets of long ribbonlike layers of turbostratic graphite. These micropores are probably the empty spaces near regions of severe distortion [5]. In the transverse plane the layer packets are rather distorted and frequently branched with individual layers extending up to distances of several tens of nanometers. The widths of the microfibrils as seen in longitudinal sections are usually less than 10 nm [5].

Thus, while a considerable amount of x-ray and electron microscopic evidence has been presented, emphasizing the microfibril or fibrillar structure for carbon fibers, it has been suggested [29] that the importance of fibrils has been overemphasized. If they do play a dominant role, one might expect the torsional rigidity to be very low, as in a "rope," but this is contrary to observations [4]. It is perhaps more useful to think of ribbons, rather than fibrils. According to Bennet and Johnson [93,94], a more realistic two-dimensional model of the three-dimensional crystallite interlinking could be represented as in Figure 17A.

Oberlin [110] has recently proposed a model for the longitudinal and transverse structure of high-modulus and high-strength PAN-based carbon fibers (Fig. 17B), on the basis of their dark-field examination of these fibers. According to these workers, the two types of carbon fibers differ only in their radius of curvature of the sheets of aromatic layers. High-modulus carbon fibers have aromatic layers with a larger transverse radius of curvature, which decreases from the periphery to the center of the fiber. The high-strength carbon fibers, on the other hand, have aromatic layers with a smaller but uniform radius of curvature.

2.3.2 Lamellar Model

Lamellar structure, although first detected by Shindo [60] in his electron diffraction studies of PAN-based carbon fibers, was also demonstrated by Johnson and co-workers [49] while studying the graphitization of PAN-based carbon fibers by examination of the fracture surfaces using scanning electron microscopy. These workers observed that carbon fibers obtained by carbonization and graphitization at 2000°C of PAN fibers, which have been partially oxidized, show a two-zone structure with a hole along the axis. When these fibers are graphitized at higher temperatures of 2260°C the material around the central zone becomes organized, with the geometry of the hole resulting in the formation of concentric lamellar graphite structures in the core of the fiber (Fig. 18). Most other parts of the fiber possess the fibrillar

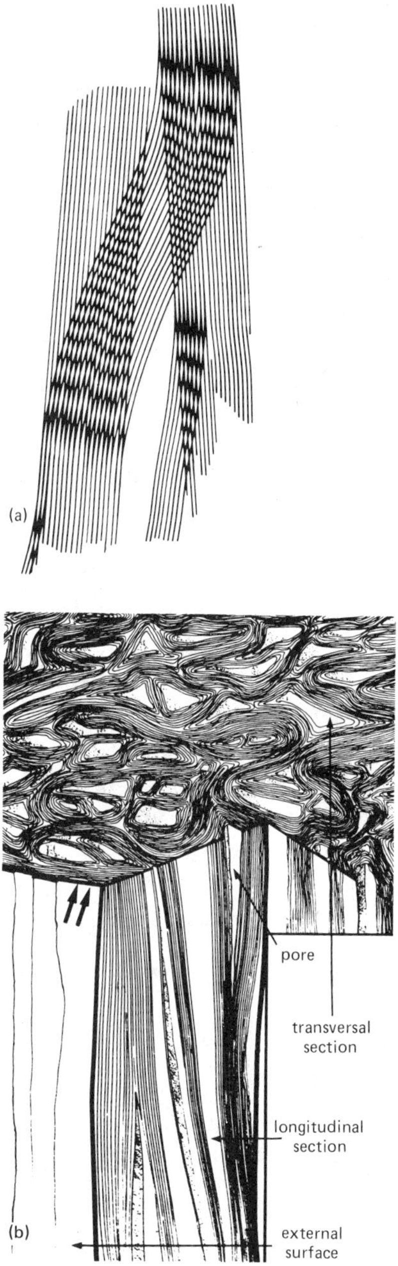

Figure 17(a) Two-dimensional representation of crystallite interlinking in carbon fiber. (From Ref. 93, reproduced with permission from Pergamon Press Inc., London.) (b) Carbon fiber model. (From Ref. 110.)

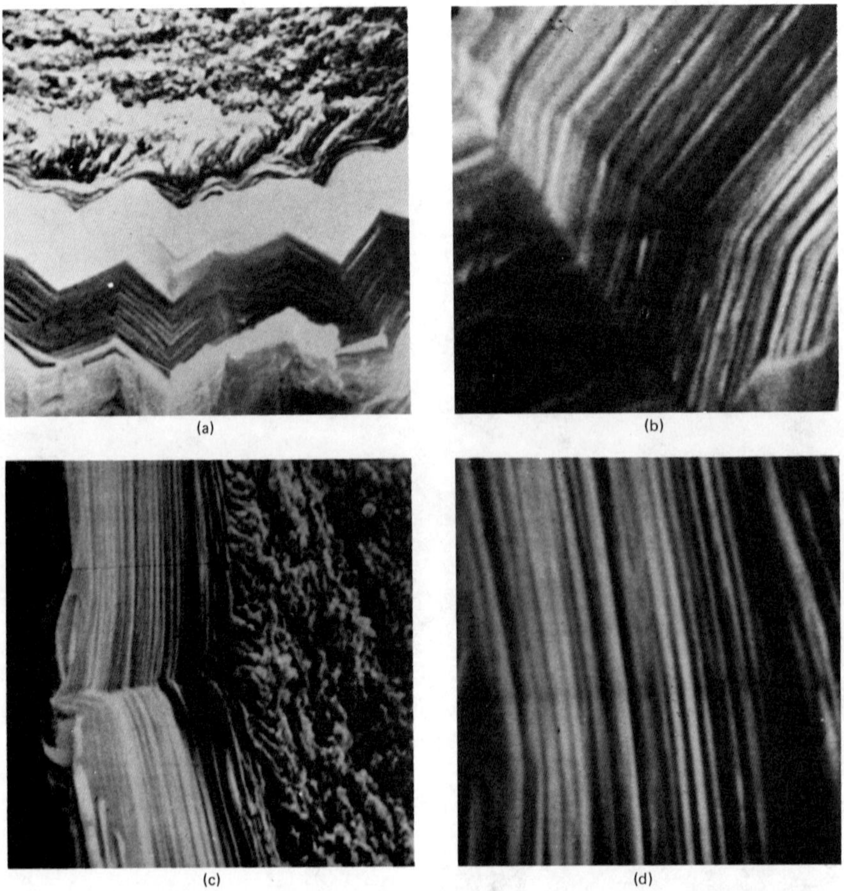

Figure 18 Lamellar structure in partially stabilized PAN-based carbon fiber. (From Ref. 49.)

structure. The formation of this lamellar structure was a function of the graphitization temperature. It developed further and extended into the entire fiber at a heat treatment temperature of 2560°C, converting some of the fibrillar structural material into a distinct lamellar texture. Since no lamellar texture was seen in the case of the fully stabilized fiber, given normal preoxidation treatment and graphitization at 2600°C [49], it was suggested that the development of the lamellar structure was almost entirely due to the partial stabilization of the fiber. The micrograph (Fig. 18) shows the high degree to which partially oxidized graphitized acrylic fibers can develop a lamellar structure.

Johnson et al. [49] suggested that partial oxidation creates an infusible oxidized crust and an unoxidized core material that passes through a fused state which is known to enhance the development of a highly lamellar structure [83]. The graphitic crystallites in the lamellar phase have grown on the surface of oriented previously oxidized phase. The lamellar phase has an L_c value of 230 nm as compared to the fibrillar structure which has an L_c value less than 10 nm.

In another study [65] of stress-graphitized carbon fibers, it was observed that these fibers often consisted of units arranged concentrically around the fiber axis, suggesting the development of a lamellar form. Although the structural development in these materials involves a mechanism different from that of partially oxidized fibers, the final structure was almost exactly similar, indicating that the lamellar structure is the ultimate stable form of the material.

Bacon and Silvaggi [47] have interpreted the transverse section of the electron micrograph of Thornel 50 rayon fiber obtained by Harling [84] as revealing a lamellar structure, consisting of packets of ribbonlike layers separated by a honeycomb of micropores.

A lamellar structure with "tree trunk" configuration was also observed by Butler and Diefendorf [44] in their examination of PAN- and rayon-based carbon fibers by polarized light microscopy, although the extent of this configuration was very small in the case of rayon fibers [5]. Since these fibers have been shown to possess a fibrillar structure, it is reasonable to believe that the lamellar structure is present along with the fibrillar structure. It is quite probable that the fibrillar structure gradually changes into the lamellar one as the preferred orientation becomes stronger [49].

Oberlin [85] and Oberlin et al. [86] have recently demonstrated that (002) lattice fringe images from longitudinal as well as from transverse thin sections of carbon fiber cannot be used to distinguish between the fibrillar or the lamellar models, because the fibrils or the folded parts of the distorted lamellae look alike in the longitudinal sections, while the continuity of the (002) fringes can result either from the real curvature of continuous layers or from a superposition of isolated details situated at different levels in the carbon fiber. It is suggested that the two structures can only be differentiated by a very careful comparison of |11| and |10| dark-field images. The dark-field high-resolution electron micrographs for pitch- and PAN-based carbon fibers are shown in Figures 19 and 20, which suggest that the lamellar model better describes the microstructure of these fibers.

Some indirect evidence in favor of the lamellar structure of carbon fibers has been provided by Donnet et al. [87] and Donnet and Ehrburger [88] in their work on the chemical properties of type I carbon fibers. It was observed that carbon fiber can be converted into

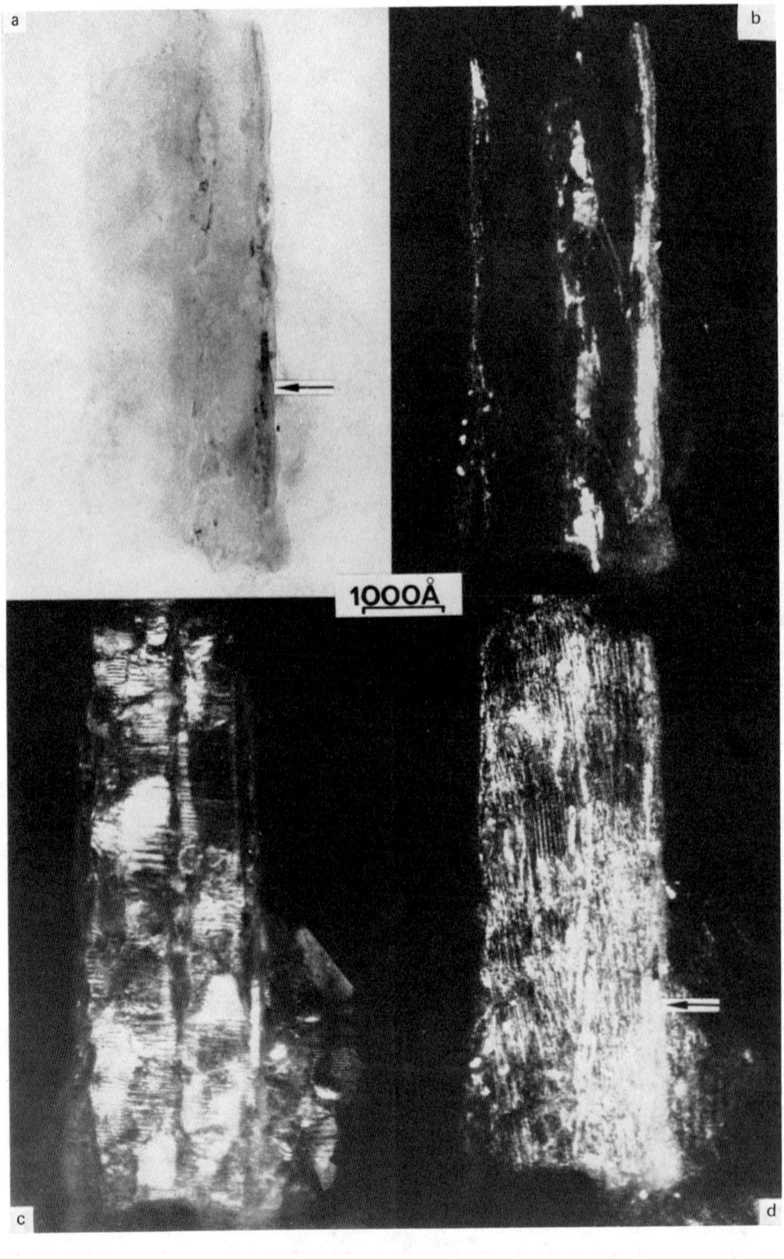

Figure 19 Bright and dark-field images of carbon fiber showing the lamellar structure. (From Ref. 85.)

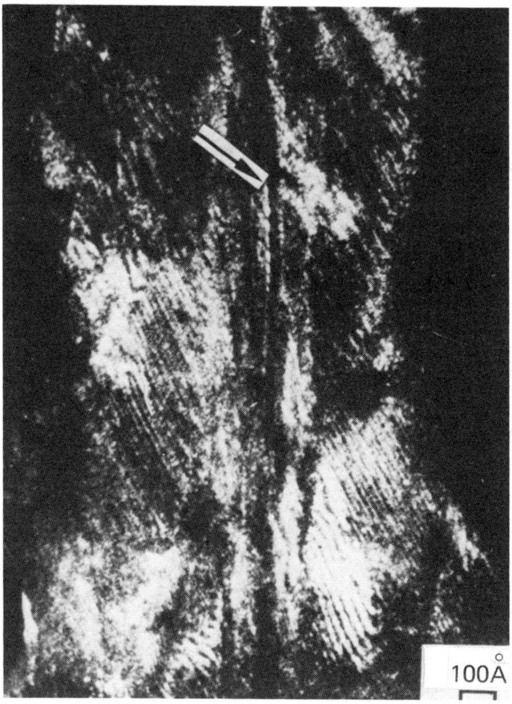

Figure 20 Dark-field image of a PAN-based carbon fiber. (From Ref. 85, reproduced with permission from Pergamon Press Inc., London.)

graphitic oxide, which is essentially a lamellar compound by treating the fibers with a mixture of H_2SO_4, $NaNO_3$, and $KMnO_4$ (Hummer's reagent) [109]. A (002) lattice fringe image of a thin fragment of a carbon fiber which has been exfoliated, as illustrated in Figure 21A, shows well ordered carbon layers in the lamellae. When heated above 200°C graphitic oxide deflagrates and graphitelike planes exfoliate. As seen in Figure 21B, this deflagration and exfoliation are very clearly observed with Hummer's reagent-treated fibers heated in the electron microscope. These experiments give strong support to the existence of graphitic layers sufficiently well ordered; graphitic oxide has been known to be able to exist only if the graphite order is good enough [105].

Thus, it seems that a carbon fiber can simultaneously possess a fibrillar as well as a lamellar fine microstructure. The fibrillar structure may, under certain conditions, gradually change into the lamellar structure. It is reasonable to believe that the lamellar structure is

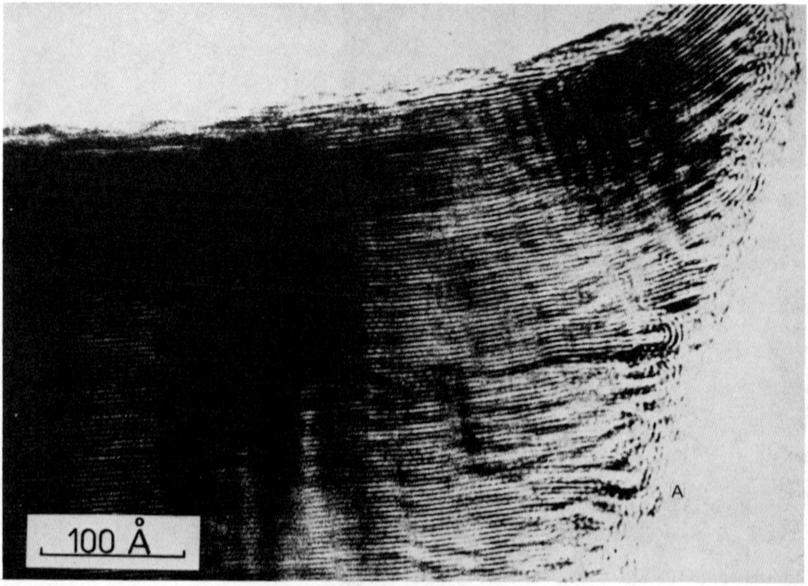

Figure 21A Lattice fringe image (002) of a thin fragment of exfoliated PAN carbon fiber showing lamellar structure. (From Ref. 87, reproduced with permission from Pergamon Press Inc., London.)

the ultimate stable form of the carbon fiber material, as pointed out by Johnson et al. [49]. In fact, the fibrillar and lamellar structures can be considered as similar, since the lamellae can be small enough to look like a microfibril, and the microfibrils can be large enough to look like lamellae.

2.3.3 Circumferential-Radial Model

The cross-sectional structure of carbon fibers, which is of importance in constructing three-dimensional structural models, has been described by several workers in terms of the circumferential radial [56] or onion-skin model [89]. Knibbs, while examining by optical microscopy end-on-end longitudinal sections of PAN carbon fibers [56] prepared under different processing conditions, identified three different types of fiber structures which differ in the orientation of the graphitic crystallites within the transverse plane. These structures are schematically represented in Figure 22. Structures a and b are two-zone structures, while structure c is a single-zone structure. Structure a has some circumferential preferred orientation of the c axis in

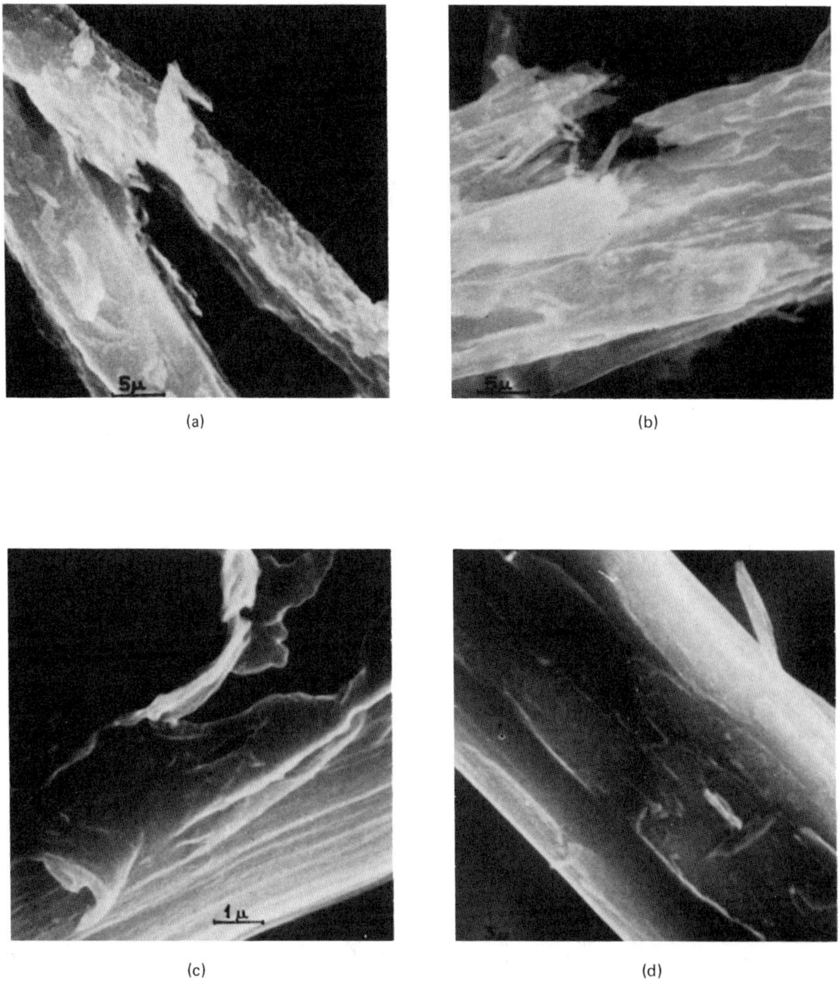

Figure 21B Carbon fiber exfoliation and deflagration obtained by rapid heating at 230°C after fiber treatment with Hummers' reagent. (From Ref. 109.)

the outer zone, while the inner zone is almost isotropic (randomly oriented). The longitudinal section of the optical micrograph, in fact, showed an isotropic central region with a noncrystalline or amorphous carbon. In the second type of structure (b), the outer layer crystallites have a radial preferred orientation whereas the central zone has a radial preferred orientation. The third type of structure, which is

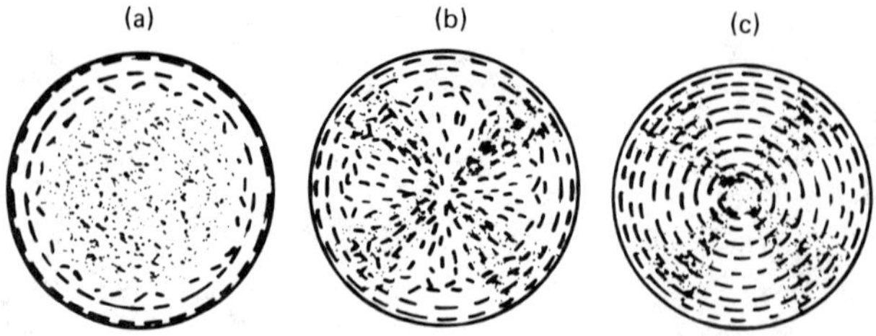

Figure 22 Schematic representation of carbon fiber structure. (a) Isotropic center and oriented outer skin, (b) different orientation in the center and outer layer, (c) one type of preferred orientation. (From Ref. 56.)

a single-zone structure (Fig. 22c), has a circumferential preferred orientation of the c axis. The existence of these three types of structures was found to depend only on the process conditions, and was independent of the heat treatment temperature, which merely enhanced the crystalline alignment without altering the direction of preferred orientation.

Diefendorf and Le Maistre [89] also observed a two-zone structure conforming to a modified "onion-skin" model with basal plane aligned parallel to the fiber surface and the degree of misalignment increasing with depth into the fiber. This configuration results in an alignment of the "c" direction with the fiber radius and the "a" direction with its circumference. The configuration of the crystallites in the case of fibers with dog bone-shaped end sections was also similar.

Barnet and Norr [90,102] examined the cross-sectional macrostructure of carbon fibers by SEM after etching in an oxygen plasma and observed that the macrostructure varied not only with the modulus of the fiber but also with its precursor, suggesting that the precursor crystallinity is important when considering the total picture of the carbon fiber structure. These workers observed that the carbon fiber consisted of a central region which was more etch-resistant, and an outer ring which was comparatively less etch-resistant: between these two regions were the radial ridges and valleys. The valleys were easily etchable and, thus, were less ordered phases of the macrostructure. This supports the "circumferential radial" model for the long-range crystalline structure in PAN carbon fibers. The structural model proposed by these workers for Modmor I is shown in Figure 23. While proposing this structure, these workers have accepted that Modmor I is an extreme case. Therefore, although the structure is thought to be fairly complete and the model incorporates all that is known or proposed regarding carbon fiber structure, all the details observed in this structure may not necessarily be seen in the struc-

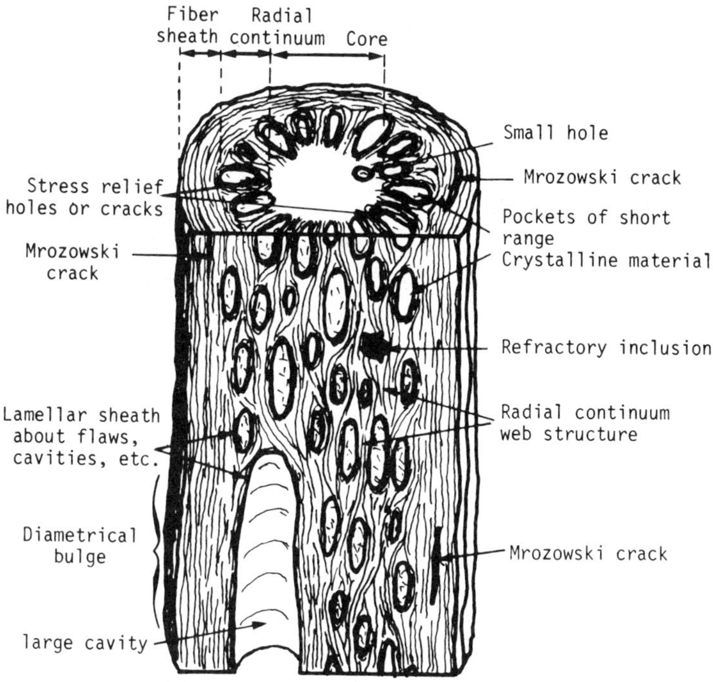

Figure 23 Structural model for a high-modulus carbon fiber. (From Ref. 102.)

ture of other carbon fibers. However, this structure may help to encourage careful study of the structural variations which may be observed in other carbon fibers.

The development of this circumferential and radial crystallinity has been used to explain the occurrence of flaws, which most commonly are found between the radial macrostructure of the fiber where a maximum degree of disorder occurs [90]. It is postulated that the well-ordered portions of the fiber resist diametrical shrinkage and produce stress fields in less ordered phases of the carbon fiber. These stress fields become stronger, due to differential thermal shrinkage between the phases upon cooling down after processing, develop flaws, and create cracks.

The radial preferred orientation texture has also been observed by Diefendorf and Tokarsky [81]. These workers are of the view that, while the texture that develops in the center of the fiber depends upon the processing conditions, an onion-skin structure is always observed on the surface.

Wicks and Coyle [55] found that the surface zones of all PAN carbon fibers have a predominant circumferential preferred orientation, although the internal structure in highly graphitized carbon fibers depends upon the diameter of the fiber. In the case of thin fibers with diameters approximately 8 μm, the orientation of the layer planes was found to be circumferential in the outer zone but radial in the interior of the fiber. In the case of thicker fibers, the configuration was more complex, with the development of a core of isotropic or randomly aligned material within the two zones, although the general axial orientation was maintained.

It may be pointed out that the observations of Wicks and Coyle may be more scientifically explained on the basis of the extent of the stabilization treatment of the fiber rather than on the diameter of the fiber. A thicker fiber is likely to have more of a partially stabilized inner zone than a thin fiber.

2.3.4 Two-Type Structure (Sheath-Core Model)

Butler and Diefendorf [44] were probably first to suggest that PAN-based type I carbon fibers prepared by heat treatment above 2000°C had a composite structure with two distinct phases. X-ray and electron diffraction studies showed that the graphitic crystallites on the surface were larger, and their basal planes generally more highly aligned parallel to the fiber axis compared to the crystallites in the interior of the fiber. The spatial orientation of the graphite basal planes indicated the presence of a tree-trunk type of configuration. The two-phase structural heterogeneity was also observed in the optical microscopic studies of polished longitudinal sections of PAN carbon fibers heat-treated to 2500°C [91]. The structure consisted of an outer zone surrounding an inner darker core. The thickness of the outer zone was a function of the time of heat-treatment in air during the preoxidation stage of the carbon fiber preparation, indicating a reaction controlled by diffusion of oxygen through the outer zone. The kinetics of the thickness of the outer zone as a function of the time of heat treatment in air showed appearance of the outer zone only after a period of one-and-a-half hours and a complete disappearance of the core after 49 hours of heat treatment in air [91]. Thus the two-phase structure is attributed to the initial preoxidation treatment, with the outer sheath formed from the fully stabilized portion of the outer region and the core from the partially stabilized inner part of the fiber.

Similar optical microscopic studies of highly graphitized carbon fibers [55,56,104] showed that they have a duplex structure consisting of an outer "sheath" of concentrically aligned layers with higher axial alignment and lower average crystallite size. X-ray diffraction studies showed that the outer sheath has a thickness of 1.5 μm and is

Figure 24 Bright-field image of longitudinal section of a carbon fiber showing (a) skin region and (b) core region. (From Ref. 94, reproduced with permission from Pergamon Press Inc., London.)

separated by a distinct annular region from the inner core which is approximately 3 μm in diameter. In the case of the carbon fibers graphitized at lower temperatures (< 2000°C), the transition between the sheath and the core regions was less distinct and the transition region occupied a larger proportion of the fiber volume.

Johnson and Crawford [103], during their examination of PAN carbon fibers by high-resolution electron microscopy, observed three distinct structural phases: a highly oriented crystalline turbostratic graphitic phase ($Z \sim 15°$) representing the sheath, a less well-oriented crystalline turbostratic graphitic phase ($Z \sim 24°$) consisting of highly interwoven fibrillar crystals enclosing many voids, resembling the less

Table 9 Structural Parameters Obtained by Electron Diffraction of Carbon Fibers Prepared from Fully Stabilized PAN Fibers

Fiber	Region	L_c (nm)	σ (%)	Z (deg)
Type I				
HTT 2500°C	Skin	10.0	0.6	14
	core	3.3	1.5	24
Type II				
HTT 1500°C	Surface	1.6	—	43
	core	1.6	—	44
Type A				
HTT 1000°C	Surface	1.2	—	43
	core	1.2	—	43

Source: Data from Ref. 93.

well-oriented core [92], and a three-dimensional graphite phase in a small proportion (Z $\sim$ 5°). Barr et al. [23] observed that this two-phase structural heterogeneity also existed in the case of mesophase pitch fibers and is introduced by the processing conditions.

Bennet and Johnson [93,94] examined thin longitudinal and transverse sections of type I (2500°C), type II (1600°C), and type III (1000°C) PAN carbon fibers, preoxidized for different time intervals before heat treatment, by dark-field and lattice-fringe electron microscopy and observed that only type I carbon fibers showed a two-phase heterogenous structure, which they termed as the skin-core structure. The skin has a variable thickness averaging 150 nm but rarely exceeding 250 nm. However, there was no distinct demarcation between the skin and the core regions, the crystallites in the two regions being interlinked through an intermediate zone. The difference in crystallite size and orientation between the skin and the core regions is evident from Figure 24. The characteristic parameters of the structure in the skin and core of a fully stabilized PAN carbon fiber are given in Table 9. The graphitic crystallites are arranged parallel to the surface in the skin and random in the core, so that the c axes of the layer planes are normal to the surface of the skin. No skin formation was noticed in type II or type III carbon fibers and in the sample which was not given preoxidation stabilization treatment before heat treatment during the preparation of the carbon fibers. It was, however, found that the thickness of the skin was independent of the time of oxidation (5 hr or more).

There was no evidence for the existence of a thick, circumferentially oriented sheath surrounding a radially oriented core in carbon fibers prepared through normal oxidative treatments, as observed by earlier workers [91,55] in their optical microscopic studies.

An examination of the transverse sections showed that the surface regions were convoluted and that there was a certain degree of folding of layers in the transverse direction, but there was no evidence for the existence of microfibrils. Structural parameters, as obtained from selected area electron diffraction (SAD) after the application of a peak resolution procedure (Table 9) clearly showed that the crystallite size is larger and the preferred orientation higher in the skin region than in the core region. These observations [6] are in conflict with those of Wicks and Coyle [55], who observed that there was a gradual decrease in preferred orientation and crystallite size from the center to the surface of the carbon fiber, as discussed earlier.

Sheath-core structure in both longitudinal and transverse sections of PAN-based type I carbon fibers was observed only in those carbon fibers which had received a nonuniform oxidation treatment [94,6]. The sheath was about 1 μm thick and had a better crystalline order than the core. Lattice fringe images from sheath and core were similar in the longitudinal section, except that the layer planes in the sheath were more extended than in the core.

Cause of Two-Type Heterogeneity

Watt and Johnson [91] found that diffusion of oxygen into the precursor fiber is a slow diffusion process and results in only partial stabilization of the interior parts of the fiber. Thus the fiber, before carbonization and high temperature treatment, has an almost completely stabilized outer region and a partially stabilized inner region. These two regions are considered to be responsible for the formation of a sheath-core heterogenous structure. This view is supported by Sprague et al. [95] and Ezekiel [96].

Johnson et al. [49] examined the morphology of fibers that had been given partial oxidation treatments and observed that the two-phase structures developed when the partially oxidized fibers were heat treated to high temperatures. These workers postulated that the core structures are associated with a melting process that occurs at low temperatures in the unstabilized core.

Layden [97] observed retrograde core formation during air oxidation of PAN homopolymer fibers. The core developed at the center of the fiber after an incubation period and grew outward, towards the circumference, as oxidation progressed. This type of retrograde core formation was absent in the case of PAN-methyl methacrylate copolymer fiber. The kinetics of the core formation appeared to be consistent with the view that the core formation is the result of a partial melting caused by the exothermic nature of the oxidation process [49].

Wicks and Coyle [55], from their optical microscopic examination, found that the thickness of the sheath in high-modulus carbon fibers

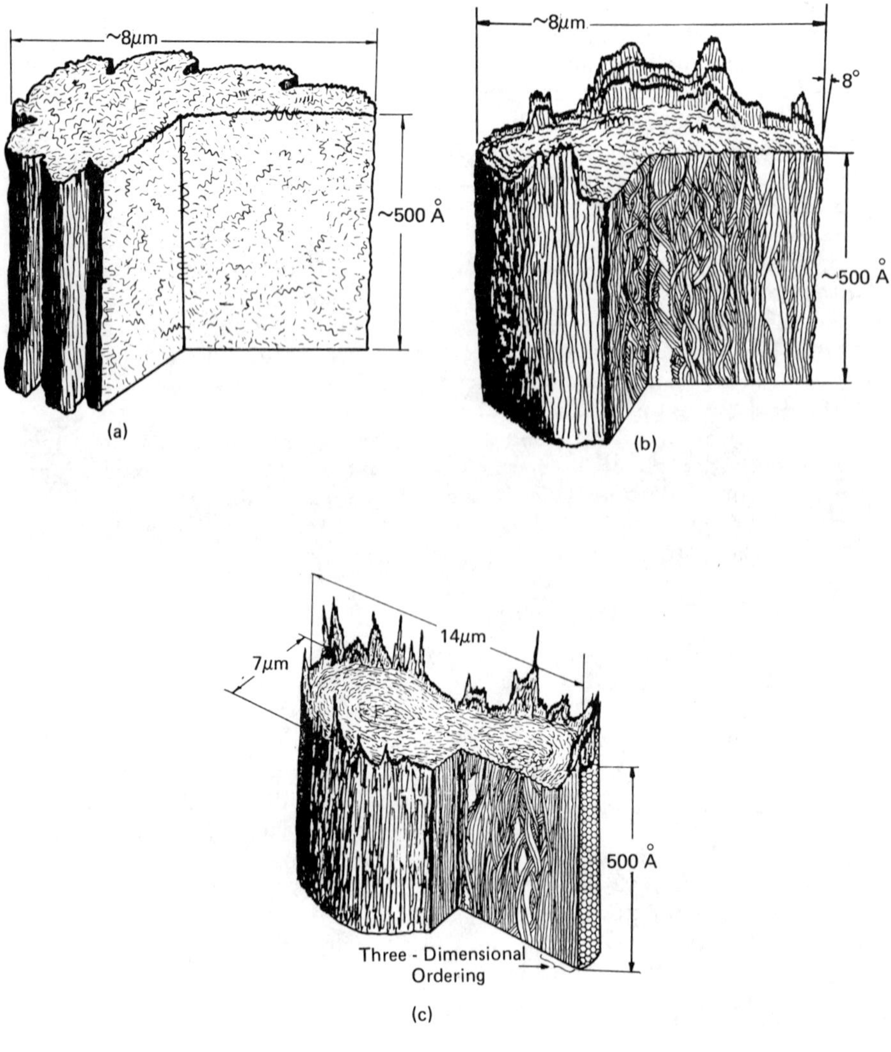

Figure 25 Schematic three-dimensional structures for carbon fibers (a) Rayon-based Wyb, (b) PAN-based Fortrail 4y, and (c) PAN-based CS-1. (From Ref. 81.)

was independent of both the initial fiber diameter and high-temperature treatment. This indicates that the cause for this two-phase structural heterogeneity is present "in latent form" in the low-temperature heat treatment and is determined by the earlier processing stages. The variation in crystalline parameters between the core and the sheath regions was attributed to the differences in structure resulting from graphitization of the fully stabilized outer region (sheath) and the formation of three-dimensionally ordered graphite within the core of the fiber resulting from fusion [49].

Sheath-core heterogeneity was observed by Johnson [6] in type I PAN carbon fibers which had received a nonuniform oxidation treatment of 6 hr in a vacuum at 230°C, followed by one-half hour in air at 230°C. This has been attributed to the fact that the sheath region in the prevacuum-treated carbon fiber is richer in oxygen [98], which enhances the lateral organization of the ladder polymer. Consequently, the subsequent heat treatment at 2500°C results in the creation of more extensive layer planes in the sheath than in the core.

2.3.5 Three-Dimensional Models

Diefendorf and Tokarsky [81] have constructed three-dimensional models for carbon fibers by combining the axial and radial structures. Three such models, which conform to optical, SEM, replica, and other electron microscopic and x-ray diffraction studies for two PAN precursor and one rayon precursor carbon fibers, are shown in Figures 25a–c. The rayon-based fiber WyB (Fig. 25a) shows an axial structure composed of relatively small microfibrils oriented at all angles to the fiber axis, but no radial preferred orientation. The PAN carbon fibers Fortrail 5y and CS-1 have complex internal radial structures, but the surface and structures are basically tree trunk and onion skin (Figs. 25b,c).

Johnson has proposed recently the model shown in Figure 26 which shows a difference in ordering and density of the carbon ribbons or microcrystallites between the peripheral zone, nearing the surface and the case of the fiber. This model illustrates the constitution of a "skin" of the fiber, which is in agreement with the behavior of carbon fibers in many processes, for example the electrochemical oxidation as pointed out by Donnet [105].

In her latest papers, A. Oberlin states that "the behavior of high tensile fibers can be interpreted and a suitable model can be drawn only if the data are compared to the results obtained on other carbonaceous materials" [86,106,107,110]. After very careful examination in transmission electron microscopy, it was concluded that high-tensile-strength PAN-based carbon fibers are made of aromatic basic structural units less than 10 Å in size associated edge-to-edge in parallel

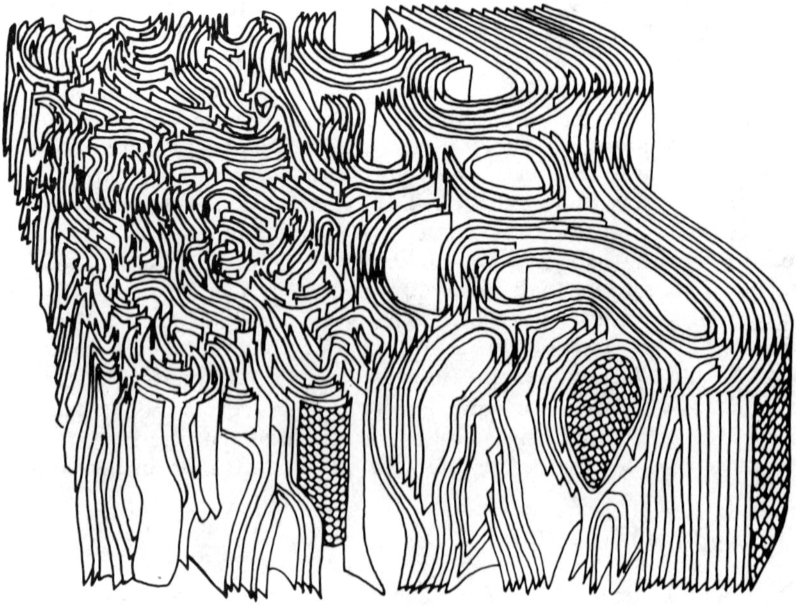

Figure 26 Carbon fiber three-dimensional model. (From Ref. 93.)

to form larger wrinkled aromatic layer sheets. These sheets are crumpled parallel to the fiber axis and then the folds are entangled. It results in a porous structure with pores elongated parallel to the fiber axes.

High-modulus carbon fibers are also described by the same model [108].

2.4 MICROSTRUCTURAL HETEROGENEITY

Carbon fibers have a heterogeneous microstructure and, consequently, several microstructural models have been suggested and discussed. Generally, these models speak of a well-oriented phase with larger crystallites and fewer voids, and a less well-oriented phase with comparatively smaller crystallite size but containing a larger number of voids. Johnson [6,111] observed the presence of a small portion of a third phase composed of well graphitized highly oriented crystallites. Similar graphitic inclusions were also observed by Donnet et al. [87] in their examination of severely oxidized PAN carbon fibers which had never been heat-treated beyond 1100°C. The high-resolution electron micrography of one such fiber is shown in Figure 27. The presence of

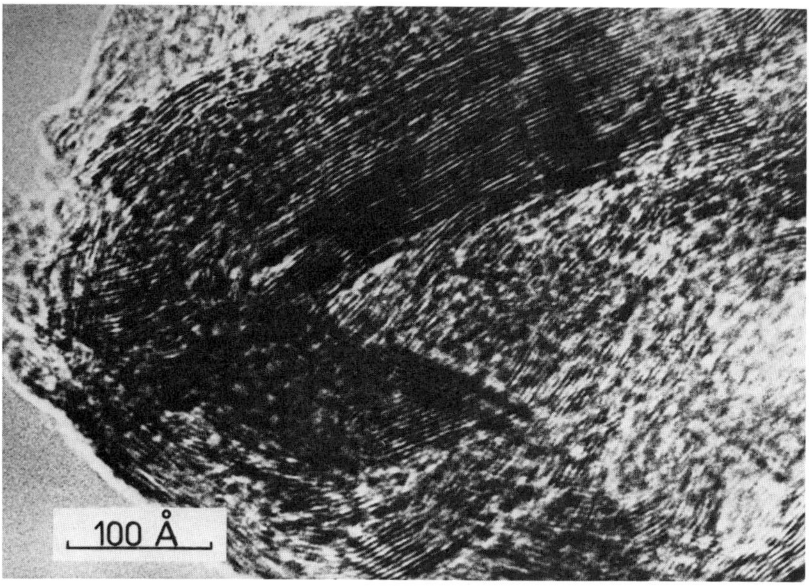

Figure 27 High-resolution electron micrographs of a severely oxidized acrylic carbon fiber (10% weight loss). (From Ref. 87, reproduced with permission from Pergamon Press Inc., London.)

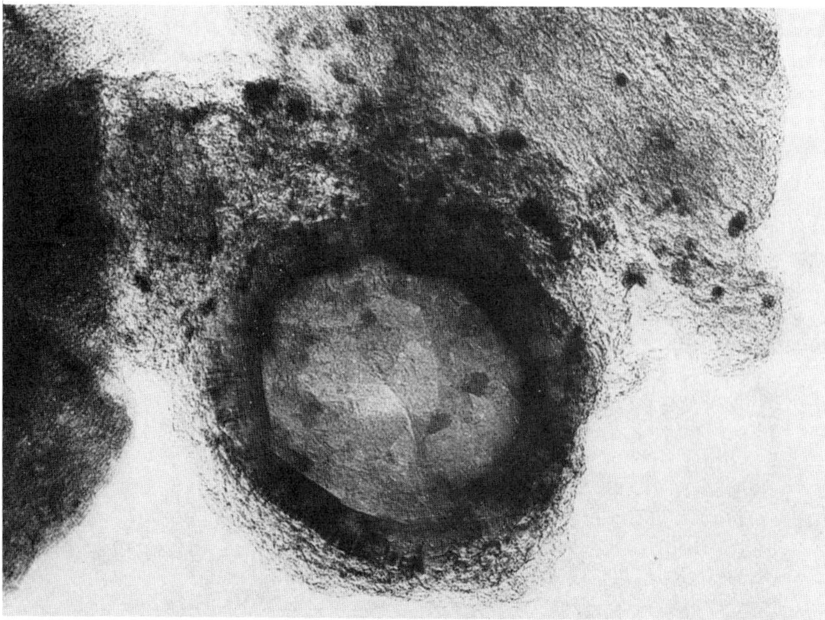

Figure 28 Highly graphitized inclusions in carbon fibers. (From Ref. 100, reproduced with permission from Pergamon Press Inc., London.)

large areas of parallel planes, which are graphitic in nature, can be clearly seen. This indicates that extensive graphitized areas exist, even in those carbon fibers which had never been exposed to temperatures beyond 1100°C. Johnson and Tomizuka [99–101] observed many well-graphitized inclusions (Fig. 28), which may be continuous or discontinuous, in the case of lignin-based and pitch-based carbon fibers heat-treated to 2000°C. The extent of these inclusions was smaller in the case of the pitch-based fibers. These inclusions have been attributed to the presence of metallic impurities, which catalyze the graphitization process. A knowledge of the presence of these heterogenous microstructures and the mechanism of their formation is of considerable technological importance in producing high-strength and high-modulus carbon fibers.

REFERENCES

1. Boder, H., Golden, D., Rose, Ph., and Wurmseher, H. Z. Werkstofftech. *11*, 275 (1980).
2. Donnet, J. B., and Voet, A., *Carbon Black*, Marcel Dekker, New York (1976).
3. McKee, D. W., and Mimeault, V. J., *Chemistry and Physics of Carbon*, vol. 8 (Edited by P. L. Walker, Jr. and P. Thrower) Marcel Dekker, New York (1973).
4. Reynold, W. N., *Chemistry and Physics of Carbon*, vol. 11 (Edited by P. L. Walker and P. Thrower) Marcel Dekker, New York (1973).
5. Bacon, R. *Chemistry and Physics of Carbon*, vol. 9 (Edited by P. L. Walker, Jr. and P. Thrower) Marcel Dekker, New York (1973).
6. Johnson, D. J. Phil. Trans. Roy. Soc. London, A *294*, 443 (1980).
7. Ergun, S. *Chemistry and Physics of Carbon*, vol. 3 (Edited by P. L. Walker, Jr.) Marcel Dekker, New York, 1968, p. 211.
8. Ruland, W. *Chemistry and Physics of Carbon*, vol. 4 (Edited by P. L. Walker, Jr.) Marcel Dekker, New York, 1968, p. 1.
9. Fourdeux, A., Perret, R., and Ruland, W. *Carbon Fibres—Their Composites and Applications*. The Plastics Institute, London, 1971, p. 57.
10. Johnson, D. J. *Carbon Fibres—Their Composites and Applications*. The Plastics Institute, London, 1971, p. 52.
11. Eastabrook and Seed, Private communication. Reference quoted No. 90 in Ref. 5.
12. Fourdeux, A., Herinckx, C., Perret, R., and Ruland, W. Compt. Rend. Acad. Sci. (Paris) *269*, 1597 (1969).

13. Fourdeux, A., Perret, R., and Ruland W. J. Appl. Cryst. *1*, 252 (1968).
14. Coyle, R. A., Gillin, L. M., and Wicks, B. J. J. Aust. Cerem. Soc. *6*, 27 (1970).
15. Ruland, W. J. Appl. Phys. *38*, 3585 (1967).
16. Yamamoto, M., and Yamada, S. Appl. Polymer Symposia No. 9, 263 (1969).
17. Bacon, G. E. J. Appl. Chem. *6*, 477 (1956).
18. Goggin, P. R., and Reynold, W. N. Phil. Mag. *16*, 317 (1967).
19. Price, R. J., and Bokros, J. C. J. Appl. Phys. *36*, 1897 (1965).
20. Gasparoux, H., Pacault, A., and Poquet, E. Carbon *3*, 65 (1965).
21. Mitchell, S. J., and Thomas, C. Carbon *9*, 253 (1971).
22. Ezekiel, M. Fibre Sci. Technol. *2*, 329 (1970).
23. Barr, J. B., Chwastiak, S., Didchenko, R., Lewis, I. C., Lewis, R. T., and Singer, L. S. Appl. Polymer Symposia No. *29*, 161 (1976).
24. Bright, A. A., and Singer, L. S. Carbon *17*, 59 (1979).
25. Warren, B. E., and Bodenstein, P. Acta Cryst. *20*, 602 (1966).
26. Ruland, W., and Tompa, H. Acta Cryst. A*24*, 93 (1968).
27. Debye, P., and Bueche, A. J. Appl. Phys. *20*, 518 (1949).
28. Porod, G. Kolloid−Z *125*, 51, 109 (1952).
29. Johnson, D. J., and Tyson, C. N. Brt. J. Appl. Phys. *D2*, 787 (1969).
30. Johnson, D. J., and Tyson, C. N. Brt. J. Appl. Phys. *D3*, 526 (1970).
31. Brydges, W. T., Badami, D. V., and Jones, J. Appl. Polymer Symposia No. 9, 255 (1969).
32. Perret, R., and Ruland, W. J. Appl. Cryst. *2*, 209 (1969).
33. Perret, R., and Ruland, W. J. Appl. Cryst. *3*, 525 (1970).
34. Uchida, T., Shinoya, I., Itoh, Y., and Nukada, K. *Tenth Bienn. Conf. on Carbon*, Lehigh University, Bethlehem, Penn., 1971, summary of papers, p. 31.
35. Perret, T., and Ruland, W. 9th Biennial. Conf. on Carbon, Boston, Mass., 1969, summary of papers, p. 158.
36. Bennet, S. C., Johnson, D. J., and Murray, R. Carbon *14*, 117 (1976).
37. Saunderson, D. H., and Windsor, C. G. *Proc. 3rd Conf. on Ind. Carbon and Graphite*, Soc. Chem. Inc., London, 1971, p. 438.
38. Sharp, R. S. J. Roy. Mic. Soc. *86*, 271 (1967).
39. Jarro, M. A. A., Ladner, W. R., and Rantell, T. D. Carbon *14*, 219 (1976).

40. Johnson, D. J. Nature *226*, 750 (1970).
41. Stewart, Murray J. Aust. Cerem. Soc. *9*, 56 (1973).
42. Hawthorne, H. M., *Carbon Fibres—Their Composites and Applications*, The Plastics Institute, Lond, 1971, p. 81.
43. Badami, D. V., Joiner, J. C., and Jones, G. A. Nature *215*, 386 (1967).
44. Butler, B. L., and Diefendorf, R. J. *9th Biennial. Conf. on Carbon*, Boson, Mass., 1969, summary of papers, p. 161.
45. Heidenreich, R. D., Hess, W. M., and Ban, L. L. J. Appl. Cryst. *1*, 1 (1968).
46. Hugo, J. A., Phillips, V. A., and Roberts, B. W. Nature *226*, 144 (1970).
47. Bacon, R., and Silvaggi, A. F. Carbon *9* 321 (1971).
48. Harris, B., Beaumont, P. W. R., and Rosen, A. J. Mater. Sci. *4*, 432 (1969).
49. Johnson, J. W., Rose, P. G., and Scott, G. *Proc. 3rd Conf. on Ind. Carbon and Graphite*, Soc. Chem. Ind., London, 1971, p. 443.
50. Thorne, D. J. *Proc. 3rd Conf. on Ind. Carbon and Graphite*, Soc. Chem. Ind., London, 1971, p. 463.
51. Moreton, R. *Carbon Fibres—Their Composites and Applications*. The Plastics Institute, London, 1971, p. 73.
52. Johnson, J. W. Appl. Polymer Symposia No. 9, 299 (1969).
53. Trent, P. E., and Bennett, R. K. *9th Bienn. Conf. on Carbon*, Boson, 1969, summary of papers, p. 160.
54. Tomizuka, I., Kurita, T., Tanaka, Y., and Watanabe, O. Yokyo Kyokai-Shi *79*, 460 (1971).
55. Wicks, B. J., and Coyle, R. A. J. Mater. Sci. *11*, 376 (1976).
56. Knibbs, R. H. J. Microsco. *94*, 273 (1971).
57. Shindo, A., and Sawada, Y. Carbon *18* 419 (1980).
58. Ruland, W. Appl. Polymer Symposia No. *9*, 293 (1969).
59. Watt, W., Phillips, L. N., and Johnson, W. Engineer (May 27, 1966).
60. Shindo, A. Report No. 317, Govt. Ind. Res. Institute, Osaka, 1961.
61. Bacon, R., and Schalaman, W. A. Appl. Polymer Symposia No. *9*, 285 (1969).
62. Bacon, R., and Tang, M. M. Carbon *2*, 221 (1964).
63. Saunderson, D. H. and Windsor, C. G. *Proc. 3rd Conf. on Ind. Carbon and Graphite*, Soc. Chem. Ind., London, 1971, p. 438.
64. Watt, W., and Johnson, W. Appl. Polymer Symposia No. *9*, 215 (1969).
65. Johnson, J. W., Majoram, J. R., and Rose, P. G. Nature *221*, 357 (1969).

66. Ruland, W. Conf. on Fibres for Composites (IPPI), Brighton, July 1969.
67. Allen, S., Cooper, G. A., Johnson, D. J., and Mayer, R. M. *Proc. 3rd Conf. on Ind. Carbon and Grphite*, Soc. Chem. Ind., London, 1971, p. 456.
68. Johnson, D. J., Crawford, D., and Oates C. *10th Bienn. Conf. on Carbon*, Lehigh University, Bethlehem, Penn., 1971, summary of papers, p. 29.
69. Shindo, A. Appl. Polymer Symposia No. *9*, 271 (1969).
70. Johnson, W., and Watt, W. Nature *215*, 384 (1967).
71. Ruland, W. J. Polymer Sci. C*28*, 143 (1969).
72. Fourdeux, A., Perret, R., and Ruland, W. C. R. Acad. Sci. Paris, Series C*271*, 1495 (1970).
73. Herinckx, C., Perret, R., and Ruland, W. Nature (London) *220*, 63 (1968).
74. Herinckx, C., Perret, R., and Ruland, W. Ber Dtsch. Keram. Ges. *46*, 400 (1969).
75. Jackson, P. W., and Marjoram, J. R. Nature *218*, 83 (1968).
76. Jackson, P. W., and Marjoram, J. R. J. Mater. Sci. *5*, 9 (1970).
77. Herinckx, C., Perrett, R., and Ruland, W. Carbon *10*, 711 (1972).
78. Voet, A., Morawski, J. C., and Donnet, J. B. Carbon *13*, 465 (1975).
79. Mencik, Z., Plummer, H. K., Jr., and Bartosiewicz, L. Carbon *13*, 417 (1975).
80. Perret, R., and Ruland, W. Kolloid-Z. U. Z. Polymere *251*, 34 (1973).
81. Diefendorf, R. J. and Tokarsky, E. Polymer Engg. Sci. *15*, 150 (1975).
82. Butler, B. L., and Diefendorf, R. J. In *Carbon Composite Technology*, Proceedings of the Tenth Annual Symposium of the New Mexico Section of ASME and University of New Mexico, College of Engineering, Albuquerque, 1970, p. 107.
83. Kipling, J. J., and Shooter, P. V. *Proc. 2nd Conf. on Ind. Carbon and Graphite* (London), Soc. Chem. Ind., 1965, p. 15.
84. Harling, D. F. *Proceedings of the 28th Annual Meeting of the Electron Microscopy Society of America*, Houston, Texas 1970 (Edited by C. S. Arceneaux) Claitor's Publishing Div., Baton Rouge, 1970, p. 468.
85. Oberlin, A. Carbon *17*, 7 (1979).
86. Oberlin, A., Molleyre, F., and Bastick, M. *Proc. 13th Bienn. Conf. on Carbon*, Irvine, Calif., 1977, extended abstracts, p. 371.

87. Donnet, J. B., Voet, A., Marsh, P. A. Ehrburger, P., and Dauksch, H. Carbon *11*, 431 (1973).
88. Donnet, J. B., and Ehrburger, P. Carbon *15*, 143 (1973).
89. Le Maistre, C. W., and Diefendorf, R. J. *10th Bienn. Conf. on Carbon*, Lehigh University, Bethlehen, Penn., 1971, summary of papers, p. 163.
90. Barnet, F. R., and Norr, M. K. Carbon *11*, 281 (1973).
91. Watt, W. and Johnson, W. *Proc. 3rd Conf. on Ind. Carbon and Graphite*, Soc. Chem. Ind., London, 1971, p. 417.
92. Jones, B. F., and Duncan, R. G. J. Mater. Sci. *6*, 289 (1971).
93. Bennett, S. C., and Johnson, D. J. *Proc. Fifth London Carbon and Graphite Conf.*, vol. 1, Soc. Chem. Ind., London, 1978, p. 377.
94. Bennett, S. C., and Johnson, D. J. Carbon *17*, 25 (1979).
95. Sprague, B. S. et al., Technical report, AFML, TR-66-161 Part II (1967).
96. Ezekiel, H. M., Technical report, AFML, TR-69-167 (1969).
97. Layden, G. K. Carbon *10*, 59 (1972).
98. Watt, W., and Johnson, W. Nature, London *257*, 210 (1975).
99. Johnson, D. J., and Tomizuka, I. *Carbon Fibres—Their Place in Modern Technology*, The Plastics Institute, London, 1974, p. 20.
100. Johnson, D. J., and Tomizuka, I. Carbon *13*, 321 (1975).
101. Johnson, D. J., and Tomizuka, I. Carbon *13*, 529 (1975).
102. Barnet, F. R., and Norr, M. K. Carbon Fibres—Their Place in Modern Technology. The Plastics Institute, London, 1974, p. 32.
103. Johnson, D. J., Carawford, D. and Jones, B. F. J. Mater. Sci. *8*, 286 (1973).
104. Wicks, B. J. J. Mater. Sci. *6*, 173 (1971).
105. Donnet, J. B. Carbon *20*, 266 (1982).
106. Oberlin, A. and Oberlin, M., Rev. Chim., *18*, 442 (1981).
107. Guignon, M., Oberlin, A., Desarmot, G. In press (private communication).
108. Guignon, M., Oberlin, A., Desarmot, G., C. R. Acad. des Sciences, in press, (private communication).
109. Donnet, J. B., Papirer, E., Dauksch, H., Carbon Fibre, Their Place in Modern Technology, London, 1974, p. 9/1.
110. Oberlin, A. Extended Abstracts 15th Carbon Conference, Philadelphia, p. 288-289 (1981).
111. Johnson, D. J. Chem. and Ind. *6*, 692 (1982).
112. Otani, S. Carbon *3*, 31 (1965).

3

Surface Properties of Carbon Fibers

Carbon fibers with excellent mechanical properties can now be produced. But the efficient translation of these outstanding mechanical properties into usable structural form such as composites has not been achieved. The composites obtained from them have poor interlaminar shear strength. This has been attributed to weak adhesion or poor bonding between the carbon fiber surface and the resin matrix. The degree of adhesion between the fiber and the binder phase, in turn, depends considerably on the quality and the state of the carbon fiber surface. A complete characterization of this carbon fiber surface is thus absolutely essential for investigations relating to the modification of carbon fiber surfaces. This requires a detailed knowledge of the amount and nature of the surface atoms and groups which are present, and which can be introduced on carbon fiber surfaces and their reactivities. It is also necessary to have knowledge of their surface area, microporous structure and pore-size distribution, surface energetics and their variation on adsorption of different gaseous and liquid species, and their internal and surface structure.

Structure having been discussed in the preceding chapter, the present chapter will deal with the surface properties, surface characterization, and the changes that can be introduced in the fiber surface by various treatments.

3.1 POROUS STRUCTURE AND SURFACE AREA OF CARBON FIBERS

Surface area is an important parameter which determines the interaction of a carbon fiber with the matrix materials and, hence, its behavior in a composite. It is a changeable parameter and can be increased by surface treatments. However, an important part of the surface area is made up of the internal surfaces of micropores. Because these pores can help in the propagation of microstructural faults and microdefects, their presence has a weakening effect on the carbon fiber strength. Consequently, a knowledge of the surface area and the porous structure, which includes the size and shape of the pores, is of fundamental significance in determining the interfacial bond and the strength of carbon fibers.

Carbon fibers are obtained by the carbonization of various polymeric materials, such as rayon, polyacrylonitrile (PAN), and certain pitches. The extent and nature of the porous structure and the surface area will, therefore, be largely dependent on the nature and morphology of the precursor polymer and the history of its heat treatment.

3.1.1 Porous Structure

The presence of pores in carbon fibers has been amply demonstrated by small-angle x-ray scattering studies of Perret and Ruland [1,3], Fourdeux et al. [2], Johnson and Tyson [4], Johnson [5], Badami et al. [6] and several other investigators [7,8]. It is believed that carbon fibers are made up of graphite basal planes in the form of microfibrils with imperfect stacking. This imperfect stacking gives rise to empty spaces between the microfibrils which form pores or voids. These pores are present in all carbon fibers although their population varies. Low-temperature carbon fibers, generally have small but numerous pores which are distributed throughout the carbonized material and become larger but less numerous when the carbon fiber is heat-treated at higher temperatures. The pores are generally needle-shaped and vary in diameter between 1 and 2 nm.

The porous structure of carbon fibers has also been investigated by density measurements, and by studying the adsorption of different gases and vapors. However, the interpretation of the data is complicated by the fact that carbon fibers have extremely fine micropores which can exhibit molecular sieve behavior. Didchenko [9] compared the surface area and pore size distribution in carbon fibers, heat-treated at different temperatures, by measuring the adsorption of nitrogen, and observed that the low-temperature carbon fiber (type II) showed a much larger total pore volume and surface area, whereas the high-temperature carbon fiber had a much larger mean pore diameter.

The development of porous structure in PAN carbon fibers, heat-treated to different temperatures, was investigated by studying the adsorption of neon and krypton and by measuring the displacement density of helium and a silicon fluid [10]. The porosity open to helium (ε) but not to silicon fluid was calculated by the equation

$$\varepsilon = 100\ \rho L\left(\frac{1}{\rho_{Si}} - \frac{1}{\rho_{He}}\right) \tag{1}$$

where ρL, ρ_{Si}, and ρ_{He} are the lump, silicon, and helium densities, respectively. It is seen (Figure 1) that the porosity increases from 1% in raw PAN fibers to about 18% when the fiber is heated at 1000°C with a minimum at 320°C. This minimum has been attributed to partial filling of the pores by the carbonization products. The maximum in

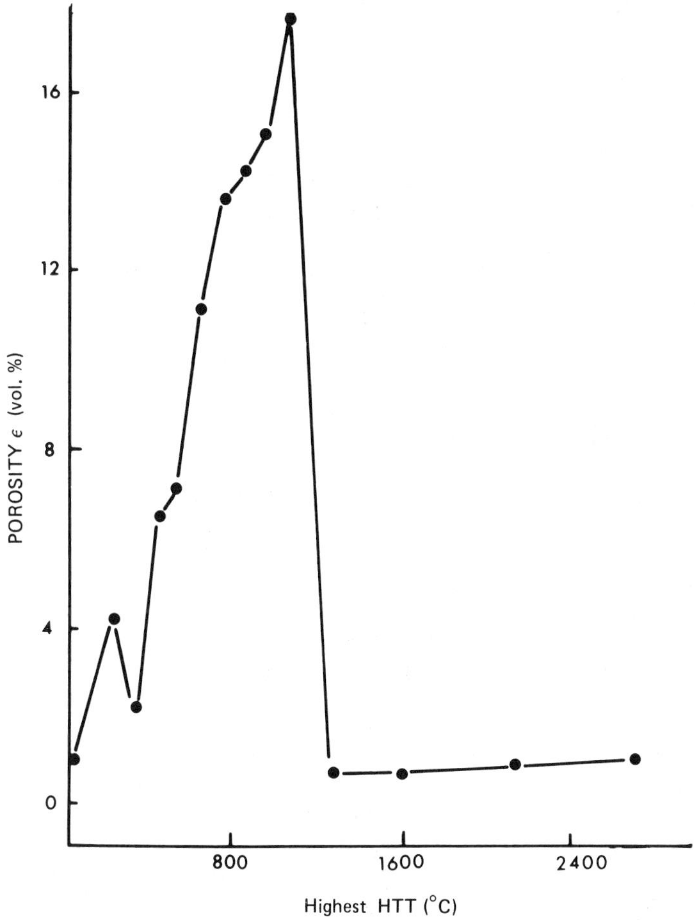

Figure 1 Variation of open porosity of PAN carbon fiber ε at 25°C as a function of heat treatment temperature. (From Ref. 10.)

the adsorption of krypton at 600°C heat treatment temperature (HTT) and the decrease thereafter [10], indicated a gradual contraction of pore entrances, producing a molecular sieve behavior. Even helium could not enter the pores above a HTT of 1200°C. This closed micropore volume was greater than 15% at 1500°C but decreased at higher heat treatment temperatures. Since silicon oil with dimensions approximating 1.5 nm could not be adsorbed, whereas helium (molecular dimensions 0.258 nm) could enter the pores, it could be inferred that the carbon fibers had pores in the range 0.258–1.5 nm.

Christner and Walker [11] measured the porosity of several cellulosic and acrylic carbon fibers, after carbonization at each 100°C interval between 320 and 830°C, using a modified molecular probe method and showed that the carbon fibers contain an interconnecting network of pores separated by constrictions. The size of the constriction opening passed through a maximum with increase in the temperature of carbonization. PAN fibers exhibited a complex carbonization behavior, and showed a double maxima in the measured porosity of fibers carbonized at increasing temperatures. The first maxima, which occurred between 330 and 430°C, indicated pore closure and fusion of the unoxidized centers of partially stabilized fibers [12]. At temperatures between 430 and 730°C the acrylic fibers underwent another pore sintering similar to cellulosic fibers. Small-angle x-ray scattering of rayon, PAN, and pitch fibers [1] showed that the pore diameter increased with increase in HTT and was independent of the starting material as well as of the chemical and mechanical treatment. The microporosity, however, was not a function of HTT but was influenced by the structure of the starting material, and the chemical and mechanical treatment.

In the case of Saran carbon fibers the micropore volume, accessible to nitrogen, decreased with increase in HTT above 1000°C. The micropores became sealed with increased HTT but were not eliminated even when the carbon fibers were heated to 2500°C [13]. The activation of the carbon fibers with CO_2 increased the diameter of the pores in the case of Saran carbon fibers [13]. In the case of acrylic fibers, however, the treatment with CO_2 resulted in the elimination of initial pores and the creation of new pores with smaller diameters [14].

The porous structure of several grades of pitch fibers was examined [15] by measuring density using a density-gradient technique and moisture adsorption at 100% relative humidity, as an indication of pore accessibility. The porous structure and fiber morphology were found to be a function of the HTT. The high-temperature fibers exhibited high density and low moisture sorption capacity indicating a low intrinsic porous structure. The low-temperature carbon fibers showed lower density and higher moisture sorption initially, but the density increased and the moisture sorption capacity decreased with increase in HTT indicating a closure of the micropores.

3.1.2 Surface Area

The surface area of carbon fibers is generally obtained by the adsorption of nitrogen or krypton at 77 K using Brunauer-Emmett-Teller (BET) equation. It depends upon the nature of the precursor material, the temperature of heat treatment, and the nature of the surface treatment given to the carbon fiber. PAN carbon fibers, generally, have uniform cross-sections and smooth surfaces and, therefore, have a

Table 1 Surface Area of Carbon Fibers Before and After Graphitization

Carbon fiber	Surface area (m^2 g)	
	Before graphitization	After graphitization
Cellulose	260	3
VyB (Union Carbide)	340	1–2

Source: Data from Ref. 16.

smaller surface area than rayon-based carbon fibers which, generally, have irregular cross-sections and rough surfaces. The carbon fibers prepared at lower temperatures (type II and type III) have less well organized structures, are highly microporous and, therefore, have high surface areas. The increase in the heat treatment temperature anneals out most of the pores, and reduces porosity open to the surface, resulting in a decrease in surface area. Bobka and Lowell [16], and Pallozzi [17], observed a decrease of two orders of magnitude in surface area on graphitization of rayon carbon fibers (Table 1).

The surface area of a carbon fiber increases considerably on oxidative surface treatments. The extent of increase depends upon the nature, the temperature, and the extent of the treatment. The influence of oxidation with nitric acid on surface area has been studied by a number of workers [18–21]. In each case the surface area increased considerably. Drzal [102,103] measured the surface area of PAN carbon fibers before and after air oxidation and observed an increase of about 10% in surface area on oxidation. This increase in surface area has been attributed to creation of micropores, in the form of surface cracks, by the oxidation treatment.

The measurement of surface areas of electrochemically etched PAN carbon fibers by nitrogen adsorption at 77 K [22,23], showed little change on oxidation. Since this etching has been shown to increase microporosity, it was suggested that the increased microporosity is not available for nitrogen adsorption. Probably the micropores are too small for the adsorption of nitrogen at this low temperature [24]. The activation of carbon fibers, by heat treatment in CO_2 in the temperature range 800–900°C, has also been found to increase the surface area [14,25].

Bansal and Chhabra [26] studied the effect of several oxidative treatments on the surface area of PAN-based carbon fibers, heat treated to 1000°C. The results (Table 2) indicate that each oxidation treatment causes an increase in surface area. This increase in sur-

Table 2 Surface Area of PAN-Based Carbon Fiber (1000°C) After Different Oxidative Treatments

Treatment	Surface area (m^2 g)	
	N_2 adsorption at 77 K	CO_2 adsorption at 273 K
Untreated	8	11
Treated with 60% HNO_3	38	45
Treated with 70% HNO_3	35	38
Treated with hydrogen peroxide	64	74
Treated with potassium persulfate	28	36

Source: Data from Ref. 26.

face area has been attributed to the formation of etch pits of molecular dimensions on the basal planes, which expose additional prismatic surfaces.

Mimeault and McKee [27] examined the influence of heat treatment in vacuum on the surface area of PAN-based Modmer II carbon fibers and observed that the surface area increased gradually with increase in the temperature of evacuation. The value, after evacuation at 1000°C, corresponded to a fourfold increase in surface area. In the case of Modmer I carbon fiber, the surface area increased but to a lesser extent and reached a constant value at about 600°C. The difference in the behavior of the two types of fibers was attributed to the history of their preparation. Modmer II carbon fibers, which had been carbonized at a lower temperature, undergo desorption of residual precursor decomposition products from the pores in the carbon fibers. Bansal and Chhabra [26] also observed a similar increase in surface area on evacuation of PAN-based carbon fibers before and after oxidation with different oxidizing agents. They have attributed this increase to the elimination of associated oxygen on evacuation resulting in the availability of additional surface which was previously occupied by the oxygen groups, and to the release of trapped carbonization products, as well as to the opening up of the microporous structure. The pore size distribution curves showed an increase both in the pore diameter and the micropore volume [44].

Turner and Dietz [28] observed that the carbon vapor deposition on the surface of PAN carbon fibers also results in an increased surface area. The surface area was almost doubled when the deposition was carried to a weight increase of about 60%.

3.2 SURFACE TREATMENT OF CARBON FIBERS

Carbon fibers, when used without any surface treatment produce composites with low interlaminar shear strength (ILSS). This has been attributed largely to weak bonding between the fiber and the matrix. Kaelble et al. [29], Larsen et al. [30], and Daukeys [31] have shown that ILSS is directly related to the fiber matrix bonding. Consequently carbon fibers were considerably inferior to boron and glass fibers which were being used in structural materials. Further, it was observed that as the temperature of graphitization of the carbon fiber increased, ILSS decreased although the Young's modulus showed an increase. These observations led investigators to develop a number of surface treatments which could improve the fiber–matrix interfacial bonding.

These surface treatments may be classified into oxidative treatments and nonoxidative treatments. The oxidative treatments can further be subdivided into gas-phase oxidations at lower or at elevated temperatures; liquid-phase oxidations carried out chemically or electrochemically, and catalytic oxidations. The nonoxidative treatments that improve the fiber–resin bonding consist in depositing more active forms of carbon, such as the highly effective whiskerization, the deposition of pyrolytic carbon, or the grafting of polymers on the carbon fiber surface, as shown on page 116.

Some average improvements in composite shear strength which can be achieved by the various surface treatments are shown in Table 3.

Table 3 Surface Treatment of Carbon Fiber and Improvement of Composite ILSS

Treatment	Improvement in ILSS (%)
1. Gaseous oxidation (air, ozone, RF plasma)	10–15
2. Liquid-phase oxidation (HNO_3, NaClO, electrolytic)	100–200
3. Whiskerization (Si_3N_4, TiO_4, SiC)	200–300
4. Pyrolytic carbon coating (CH_4, FeC, SiC)	60–100
5. Polymer grafting	80–100

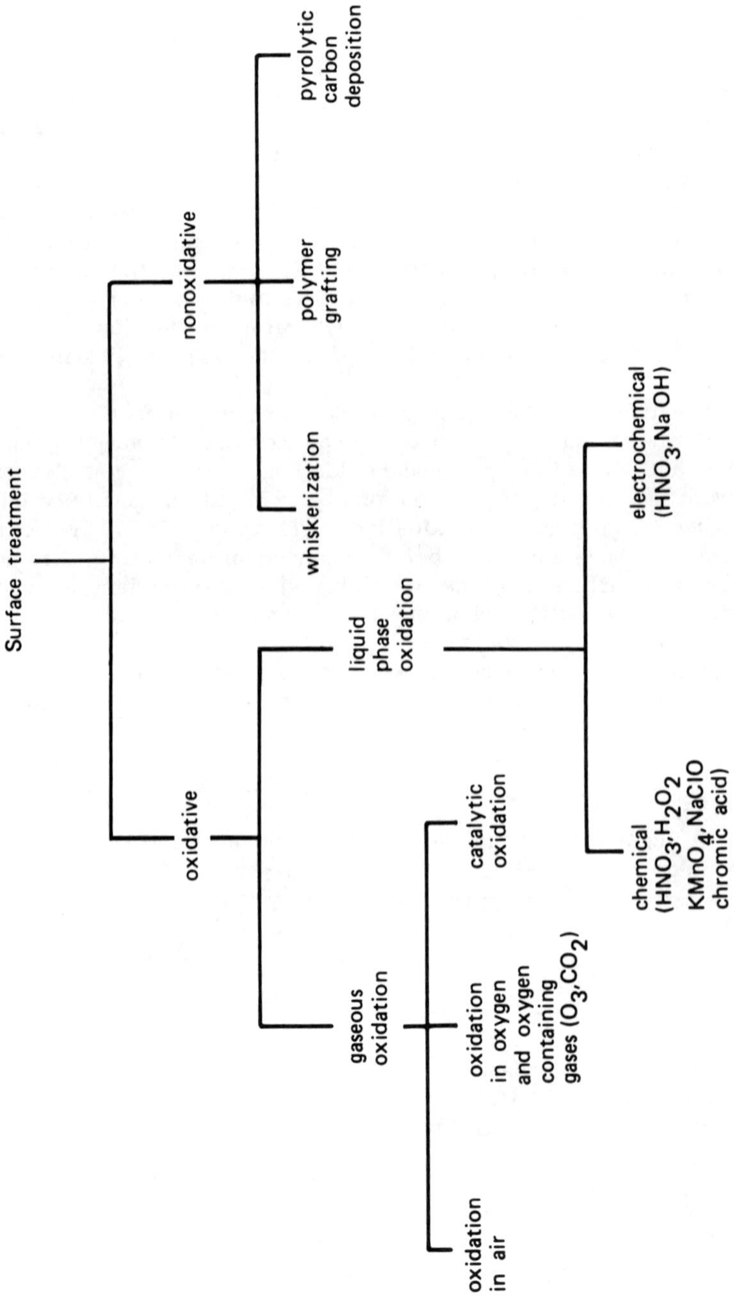
Surface treatment
nonoxidative
whiskerization
polymer grafting
pyrolytic carbon deposition
oxidative
gaseous oxidation
oxidation in air
oxidation in oxygen and oxygen containing gases (O_3,CO_2)
catalytic oxidation
liquid phase oxidation
chemical (HNO_3,H_2O_2 $KMnO_4$,NaClO chromic acid)
electrochemical (HNO_3,Na OH)

3.2.1 Gas-Phase Oxidative Treatments

The gas-phase oxidative treatments are carried out with air, oxygen, or oxygen-containing gases such as ozone and CO_2. The oxidation may be carried out either as such or in the presence of a catalyst. Air is a very commonly used gaseous oxidizing agent, which is sometimes carried out in the presence of certain other gaseous species, such as CO_2, CO, or water vapor. These species tend to modify the oxidation behavior. The gaseous treatments may be carried out at low temperatures or at eleveated temperatures. The treatments at elevated temperatures are, however, generally very drastic and cause severe degradation and excessive pitting of the carbon fiber surface, which reduces the fiber strength. Therefore, these are generally avoided. Treatment with ozone at temperatures between 100 and 150°C or with plasma-excited oxygen atoms at room temperature are milder and do not produce excessive degradation and etch pitting of the carbon fiber.

Oxidative Treatment in Air

This treatment, although very convenient, has not been very efficient in improving upon the composite shear properties. In order to obtain good results, the treatment has to be carried out at elevated temperatures where degradation and a large weight loss are likely to occur. Herrick et al. [18], for example, observed little or no improvement in the composite shear properties of rayon-based carbon fibers on treatment in air for 16 hr at 500°C. But the treatment for the same period at 600°C improved the composite shear strength by about 45%. This temperature is very close to the ignition temperature of the carbon fiber and, therefore, the treatment would result in excessive weight loss. Goan and Prosen [19], however, observed an increase in shear strength for Thornel 25 and Thornel 40 carbon fibers after treatment in air at 400°C. The treatment resulted in an increase in surface area and surface functionality of the fiber [18,19].

The isothermal oxidation of PAN, rayon, and mesophase pitch carbon fibers in air, between 230 and 315°C [32], showed that pitch- and rayon-based carbon fibers with comparable modulus showed similar oxidation rates but they were more oxidation-resistant than PAN fibers. Oxidation of rayon fibers with air—nitrogen mixtures at 1200°C [20], increased the adsorption of cationic species, but decreased those of anionic species. The reactivity data showed that the treatment resulted in the fixation of appreciable amounts of oxygen in the form of functional groups on the surface. The oxygen also intercalated between the graphitic lattices.

The effect of oxidation in air on the mechanical properties of carbon fibers has been variously interpreted, and has been shown to depend upon the experimental conditions under which the oxidation is carried out. At elevated temperatures the oxidation causes marked pitting of the carbon fiber surface which reduces strength [33]. McKee and Mimeault [21], during their examination by SEM of Modmer I and Modmer II carbon fibers oxidized in air at 700°C, observed that the etch pits were produced only during the initial stages of the oxidation process. Continued oxidation resulted in the elimination of concentric layers of carbon, rendering the surface more and more smooth and uniform with a considerable decrease in the diameter of the carbon fiber. Transmission electron microscopic (TEM) surface replicas of type I and type II carbon fibers, oxidized between 400 and 600°C, showed that the pits formed in lines which coalesce into channels [34]. The points of attack were parallel to the fiber axis at places where the carbon fibers are aligned approximately parallel to the fiber axis in such a way that the edge planes rather than basal planes and the imperfections such as point defects in basal planes are exposed. The measurements of surface area by BET (N_2) showed a rapid increase with burnoff followed by a slow decrease. The rapid increase is due to the formation of pits and the slow decrease to the formation of channels by the coalescing of the pits. Interlaminar shear strength of the fiber composites increased on oxidation, which has been attributed partly to an increase in surface area and partly to the formation of surface groups.

Sach and Bromley [35] heat-treated carbon fibers in air containing small amounts of several oxidation inhibitors such as halogens, SO_2, and halogenated hydrocarbon (CCl_4), and found that the oxidation behavior was considerably modified. The oxidation was slower, milder, and did not cause excessive pitting of the fiber surface. As a result, the adhesion between the carbon fiber and the resin matrix was considerably enhanced. McKee and Mimeault [21], however, observed no such inhibition of the oxidation rate during their oxidation treatments of Modmer carbon fibers with air in the presence of chloroform or carbon tetrachloride. Certain phosphorous compounds are well known as oxidation inhibitors for graphite, but so far they have not been used in carbon fiber treatments.

Oxidative Treatment in Oxygen

There has been limited use of oxygen for the oxidative treatment of carbon fibers because the treatment has to be carried out at comparatively higher temperatures (> 400°C), where the probability of the spontaneous ignition of the carbon fiber is quite large. Thermograms for Modmer type I and type II carbon fibers, heated at a heating rate of 5°C min^{-1}, show that both types of carbon fibers have ignition temperatures below 700°C [21], although type II carbon fiber ignites at considerably lower temperatures. This is probably due to the presence of more disorganized carbon in this fiber, less organized carbon being more reactive as shown by Donnet [133].

Bobka and Lowell [16] observed that oxidation of graphite yarn in oxygen at 400–600°C was nonuniform and resulted in the oxidation of only a small portion (≅5%) of the fiber surface. Goan and Prosen [19] oxidized Thornel 25 in oxygen at lower pressures in the presence of an rf discharge and observed an improvement in the shear strength.

Molleyre and Bastick [14] carried out the oxidation of PAN and rayon carbon fibers in oxygen at 550°C, using thermogravimetry and gasometry, and examined the internal as well as the external structure of the treated and untreated carbon fibers, using adsorption and SEM techniques. The oxidative treatment enhanced the surface area, surface rugosity, and pore diameter. The increase in rugosity caused an increase in fiber-resin composite shear strength.

Several oxygen-containing gaseous mixtures and oxygen-containing gases, such as CO_2 and ozone, have also been used for oxidative treatment of carbon fibers. Scola and Basche [36], Druin et al. [37], and Cunningham [38] oxidized carbon fibers in an inert atmosphere containing small amounts of oxygen and observed that the oxidation increased the specific surface area and the ILSS of the fiber-resin composite produced by using the modified fiber.

Adams and Boucher [25] heat-treated Saran carbon fibers in CO_2 at 900°C to different weight losses. The carbon fibers revealed an irregular oxidative attack, probably due to oxidation catalyzed by residual sodium ions [39–41]. The carbon fibers prepared in air showed a slightly more enhanced oxidation than those prepared in nitrogen atmosphere. The treatment of PAN carbon fibers with CO_2 was found to be markedly different from the treatment in oxygen [14]. The treatment in carbon dioxide smoothed the initial pores and produced new pores with smaller dimensions, whereas the treatment in oxygen increased the pore diameters. In addition a smoothing of the fiber surface was observed on treatment with carbon dioxide at 850°C [14].

Goan and Prosen [19], and Kalnin [42] oxidized carbon fibers in a mixture of oxygen and ozone at lower temperatures and observed an appreciable increase in the ILSS of the carbon fiber composite. Bansal and Chhabra [26,43,44] oxidized PAN-based carbon fibers (HTT 1000°C), suspended in water, by bubbling ozonized oxygen containing about 3% ozone, for a period of 6 hr and observed that appreciable amounts of CO_2 were produced during the oxidation treatment. The total weight loss during the treatment was about 6%. The treatment resulted in the chemisorption of appreciable amounts of oxygen and increased the surface area of the carbon fiber. The treatment, although mild and not inducing an excessive burnoff, was quite effective in the development of acidic functional groups on the fiber surface which are known to enhance the composite properties.

Catalytic Oxidation

Oxidation of graphite has been found to be greatly enhanced by the presence of small amounts of metallic impurities such as oxides of Cu,

Pb, V, and transition metals [45,46]. The presence of these metal oxides causes rapid pitting of the entire graphite surface at considerably lower temperatures. Thus, McKee and Mimeault [21] treated Modmor carbon fibers with solutions of copper acetate and lead acetate in order to incorporate metallic impurities into the fiber surface. The carbon fibers so treated were oxidized in air and composites were prepared from the oxidized carbon fibers. The composites showed considerable improvement in their shear properties, indicating that mild catalytic oxidations may be better than the more drastic gaseous oxidations. However, suitable conditions for optimum improvement in composite properties have not been published.

3.2.2 Liquid-Phase Oxidative Treatments

Solution-phase oxidation of carbon fibers has been found to be quite effective in improving the shear properties of carbon fiber-resin composites. Several types of oxidizing agents, such as nitric acid, acidic potassium permanganate, acidic potassium dichromate, sodium hypochlorite, hydrogen peroxide, and potassium persulfate have been used with varying degrees of success. These liquid-phase oxidative treatments are milder compared to the gaseous-phase oxidative treatments and generally do not cause excessive pitting and degradation of the carbon fiber. Consequently, these are preferred and efforts are being made to make these treatments more efficient and effective for maximum improvement in shear properties.

Oxidative Treatment with Nitric Acid

This is the most widely used wet oxidation treatment and has been studied extensively. Several of the factors which are known to improve the shear properties of the composites are based on studies of oxidative treatment with nitric acid. Many of the parameters, such as the concentration of the acid, the time, and the mode of treatment are now well known for optimizing the composite shear properties.

Herrick [47] and Herrick et al. [18] studied the effect of concentration of nitric acid and the time of treatment under reflux on the oxidation of WyB and Thornel type I carbon fibers, and observed that both the surface area and the surface functionality increased on oxidation. The development of surface functionality by treatment at boiling temperature with different concentrations of nitric acid as a function of time is shown in Figure 2. Concentrations less than 60% do not appear to be sufficiently effective. It is also evident that in the case of treatment with 70% nitric acid, although the development of surface functionality is poor (see Table 4), the increase in surface area is quite considerable. The effect of oxidation on the development of surface functionality and surface area is more pronounced in the case of WyB than in the

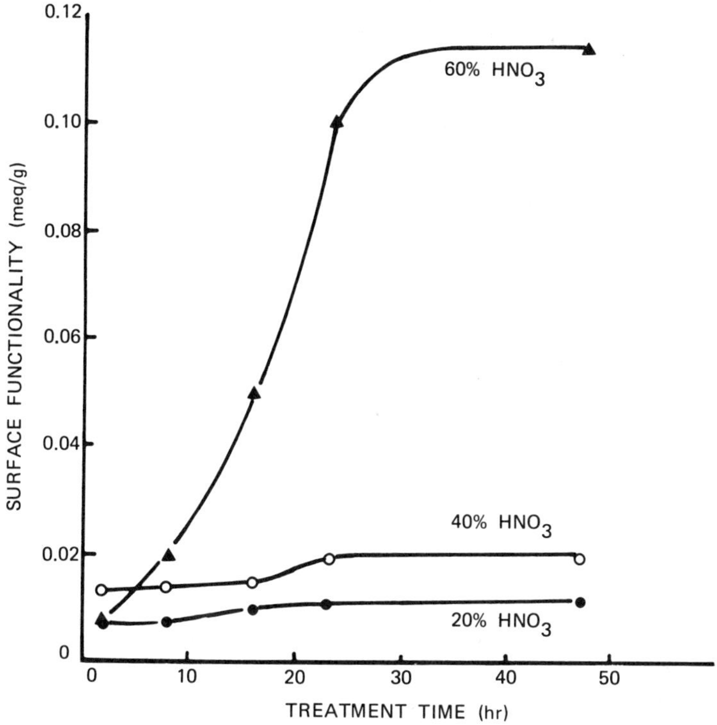

Figure 2 Development of surface functionality on treatment of carbon fiber (WyB) with different concentrations of nitric acid. (From Ref. 21.)

case of Thornel carbon fiber. Scola and Brooks [48] also observed that the nitric acid treatment caused an increase in surface area and the amount of surface oxygen complex on unit weight basis, but did not observe any significant change in the surface concentration of the latter.

McKee and Mimeault [21] oxidized Modmor I and Modmor II by refluxing with 60% nitric acid for varying intervals of time. The oxidized carbon fibers were washed with boiling distilled water, air dried at 110°C, and then vacuum dried at 50°C for several hours. The fibers differed widely in their susceptibility to oxidation by nitric acid, depending upon the temperature of their carbonization and graphitization. Modmor I carbon fibers which had been heat-treated to 2800°C lost about 4% of their weight during the oxidation treatment and remained unaffected during the subsequent treatments. Modmor II which had been heated to 1200°C, on the other hand, showed a

Table 4 Development of Surface Functionality and Surface Area on Treatment of Carbon Fibers with Nitric Acid

Fiber	Treatment	Functionality (mEq/g)	Surface area (m^2 g)
WYB 85–1/2[a]	None	—	2
	60% HNO_3	0.25	338
	70% HNO_3	0.10	126
Thornel 25[a]	None	0.002	1
	60% HNO_3	0.005	6
	70% HNO_3	0.006	12
PAN[b]	None	—	8
	60% HNO_3	—	38
	70%	—	35

Source: [a]Data from Ref. 18.
[b]Data from Ref. 26.

weight loss of about 50% after 72 hr boiling with nitric acid. Scanning electron microscopy (SEM) examination did not reveal any appreciable microscopic difference in Modmor I after the oxidation treatment, whereas Modmor II showed significant variations. The diameter of the fiber decreased from 8 μm to 5.5 μm and the rugosity was eliminated. Both the surface area and the surface functionality increased on oxidation in both fibers.

Fitzer and co-workers [49,50] and Cziollek et al. [51,51a] also oxidized type I and type II PAN carbon fibers by refluxing with 65% nitric acid, followed by extraction in water and drying at 150°C for 48 hr. Both types showed an appreciable increase in their surface oxide concentration, which increased with increasing time of treatment. The total surface oxide concentration of the high-modulus carbon fiber treated for 36 hr with nitric acid was only one-fourth (1.5 weight %) compared to the type II fiber after 24 hr oxidation (6.5 w %). The thermogravimetric analysis of PAN carbon fibers before and after oxidation with nitric acid showed formation of appreciable amounts of oxygen-containing surface groups on the oxidized samples [52].

Bahl et al. [53] treated PAN carbon fibers with nitric acid under varying conditions of concentration of nitric acid and time of treatment, and observed that 50% concentration of nitric acid and 50 minutes of time of treatment were optimum conditions for improving strength of carbon fibers. The treatment resulted in appreciable smoothing of the fiber surface by removing surface irregularities and in the introduction of appreciable amounts of acid functional groups. The amount of these surface groups increased with increase in the concentration of nitric acid or the time of treatment. These workers have postulated that the oxidation initially eliminates the less organized portions of the fiber, resulting in surface smoothing. But as the oxidation is continued, it causes removal of some of the more organized portions as well, and thus lowers the strength of the fiber.

Donnet [54] and Donnet et al. [55] observed that the treatment of type I and type II acrylic carbon fibers with boiling 68% nitric acid resulted in a significant increase in the oxygenated group content in both cases. The weight loss after 24 hr treatment was 32% in type II, and 0.1% in type I carbon fibers. The specific surface of the type I carbon fiber, however, increased markedly.

Donnet and co-workers have also confirmed the smoothing of the surface by nitric acid treatment, this smoothing being more pronounced in the case of type II fibers [134,135].

Type I carbon fibers were also treated with Hummers' reagent (which is a mixture of $NaNO_3$, H_2SO_4, and $KMnO_4$). This treatment caused appreciable chemisorption of oxygen with formation of functional groups such as $-COOH$ and $-OH$, identified by infrared (IR) spectroscopy. The oxygen content of the fiber surface increased with increase in time and temperature of treatment. The surface area, however, remained almost unchanged. The high oxygen pick up is mostly due to the formation of graphitic oxide between the graphitic layers, this graphitic oxide having been shown by IR and ESCA measurements.

Rand and Robinson [56,57], however, observed a marked increase in the external surface area after oxidation of PAN-based high-modulus carbon fibers. The active surface area, as determined by oxygen chemisorption at 340°C, increased 3–30 times indicating a considerable increase in the fraction of the surface displaying edge sites. The oxygen was picked up by the carbon fibers in the form of functional groups such as carboxyl and hydroxyl. Brooks et al. [20] also observed that the oxidation of rayon carbon fibers with nitric acid resulted in the chemisorption of oxygen in the form of functional groups.

Bansal and Chabra [26] and Bansal et al. [58] oxidized PAN-based carbon fiber (HTT 1000°C) by heating with nitric acid (60–70%, 270 ml/g) on a water bath at 90 ± 2°C. When about 10 ml of the acid was left behind, the contents were diluted with distilled water and filtered. The fiber sample was washed with distilled water until free of nitrate ions and dried at 110°C. The carbon fiber underwent a burnoff between 13 and 27% depending upon the concentration of the nitric acid. Treatment with 60% nitric acid resulted in the chemisorption of a larger amount of oxygen but a smaller amount of burnoff than the treatment in 70% nitric acid. This has been attributed to the loss of part of the active sites during the excessive burnoff of the carbon fiber surface on treatment with higher concentrations of nitric acid. The treatment was found to increase the surface area of the carbon fiber considerably [26,138].

Other Wet Oxidation Treatments

Several other oxidizing solutions with varying oxidizing power have been used for the oxidation of carbon fibers, but the details of many of these treatments are a part of the patent literature. Well and Colclough [59], Druin and Dix [60], and Scola and Brooks [48] oxidized carbon fibers by contacting with aqueous solutions of hypochlorites at moderate temperatures, and observed that the surface characteristics of the treated carbon fibers were considerably modified. The carbon fibers picked up appreciable amounts of oxygen, which resulted in enhancing the carbon-resin interfacial bond [59] and improving the adhesion of the fiber with the resin matrix [34,60], producing composites with better shear strengths. The treatment with hypochlorites is milder and inhibits excessive etch-pit formation [48].

Oxidation with aqueous solutions containing sodium dichromate and sulfuric acid, for short intervals of time [61], has also been found to improve the adhesion of the fiber to the resin matrix. Didchenko [9], however, found that the treatment of Thornel fibers with chromic acid cleaning solution for 5 min at 50°C resulted in a loss of surface area and the production of pores with larger diameters. These changes were accompanied by a marked decrease in tensile strength probably due to the excessive degradation of the carbon fiber.

Treatment of a graphite yarn in an aqueous solution containing potassium permanganate and sulfuric acid caused fixation of appreciable amounts of oxygen on the surface of the fiber and enhanced the surface area of the fiber [16]. The increase depended upon the temperature and the time of treatment. Under optimum conditions of time and temperature, the composite shear strength and the compressive strength could be doubled. This treatment for short intervals of time was found to be even more effective than the nitric acid treatment in introducing carboxyl groups into rayon carbon fibers [18].

Bansal and Chhbra [26] and Bansal et al [58] oxidized PAN-based carbon fibers (HTT 1000°C) with aqueous solutions of mild oxidizing agents such as hydrogen peroxide and potassium persulfate by contacting the carbon fiber with their aqueous solutions in closed bottles for periods up to 24 hr. Both treatments resulted in the chemisorption of appreciable amounts of oxygen, the amount depending upon the capacity of the treatment to cause a burnoff of the carbon fiber. The burnoff of the sample during these treatments varied between 6 and 12%, indicating that the treatments were milder and did not cause excessive etch-pitting and degradation of the fiber, but could introduce appreciable amounts of oxygen groups. The wettability of the fibers showed an increase both towards organic, inorganic, and polymeric liquids.

Anodic Oxidation of Carbon Fibers

The possibility of oxidation of carbon fibers in electrolytic baths of acid and alkaline aqueous solutions has been indicated [62,63] and industrially used. This treatment is preferred because it can allow continuous processing of the carbon fiber.

The pioneer work in the field was carried out by Donnet et al. [23,64] who oxidized type I (2500°C) and type II (1100°C) acrylic carbon fibers anodically in solutions of dilute nitric acid (10% by weight) containing 2.5% potassium dichromate and sodium hydroxide (1 N), using a graphite plate as the cathode. The work potential was ≅ 2 volt and a current density close to 1 mA cm^{-2}. Oxidation in alkaline medium did not appreciably change the oxygen content of the carbon fibers although oxidation in nitric acid solution increased it considerably. The treatment of type I carbon fibers was accompanied by the formation of graphitic oxide, while no graphitic oxide was noticed in the case of the type II fibers. However, the type II fibers underwent appreciable weight loss due to the evolution of CO_2. In addition, small amounts of polycarboxylic acids, which are degradation products, were also formed. The topography of the treated fibers by SEM showed a smoothing of the surface by anodic etching both for acidic and alkaline treatments. No appreciable change in surface area was noticed as a result of the treatment. Composite shear strength did not show any appreciable improvement when nonpurified carbon fibers were used. However, the shear properties showed significant increase when the composites were prepared by using the purified (extracted) fiber (Table 5). This has been attributed to the presence

Table 5 Influence of Degradation Products on the Shear Strength of Composites

	σ_c (MPa)	
Fibers	Nonpurified fiber	Purified fiber
AC nontreated		55
AC HNO_3 1 min	53	55
AC HNO_3 5 min	49	70
AC HNO_3 10 min	48	68

AC means acrylic carbon fiber heat treated to 1100°C.
Source: Data from Ref. 23.

of degradation products in the case of nonpurified fibers, which tend to decrease cohesion at the fiber-resin interface. Fitzer [138] has shown recently that the removal of these degradation products by washing the surface with alkaline aqueous solution markedly improves the performance of the heated fiber. The efficiency of this washing process was earlier demonstrated by Donnet and co-workers [136,137].

3.3 INFLUENCE OF OXIDATIVE TREATMENTS ON COMPOSITE PROPERTIES

Herrick et al. [18] observed that the oxidation of rayon carbon fibers with nitric acid resulted in an increase in the shear strength. Since the nitric acid treatment causes an increase both in surface area and surface functionality, the increase in shear strength may be attributed to both. But when the oxidized carbon fiber was treated with hydrogen—a treatment which eliminates surface functionality, but does not appreciably affect the surface area—the shear strength of the composite decreased. This shows clearly that it is the surface

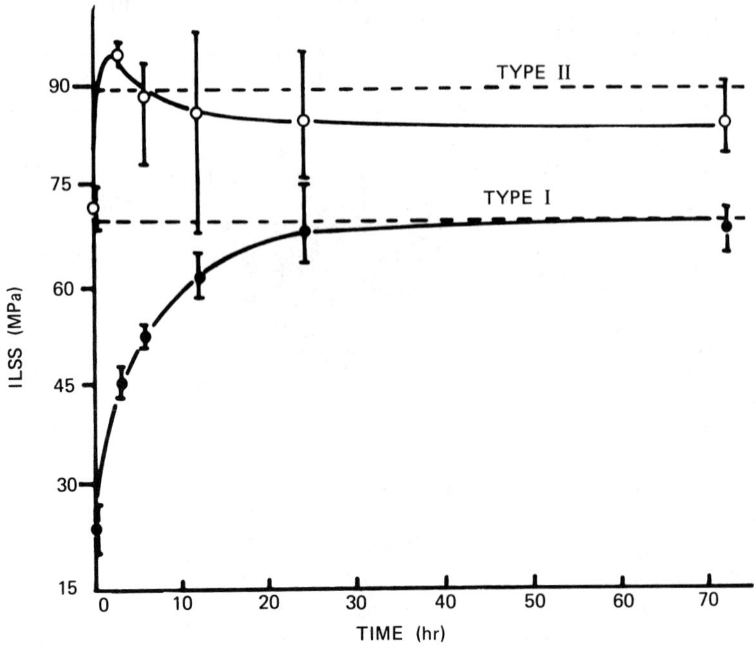

Figure 3 Effect of time of treatment with nitric acid on ILSS of type I and type II carbon fibers. (From Ref. 21.)

Table 6 Effect of Fiber-Surface Area on Composite ILSS

Fiber	Treatment	Surface area (m^2 g)	Interlaminar shear strength MPa
High-modulus Modmer	None	0.38	23
	Outgased at 900°C	0.62	29
	60% HNO_3 72 h at 118°C	1.64	71
	60% HNO_3 72 h AT 118°C outgased at 900°C	4.26	70
High-strength Modmer	Supplier's treatment	0.24	92
	Supplier's treatment, outgased at 900°C	0.92	85

Source: Data from Ref. 21.

functionality which is responsible for the increase in shear strength. Brooks et al. [20], however, contested this view and suggested that the relative importance of surface area and surface functionality in enhancing the shear strength is difficult to assess. Although oxidation with nitric acid may increase the amount of surface complexes on unit weight basis, the surface concentration of the groups may not vary significantly because the treatment is accompanied by an increase in surface area. Furthermore, these workers observed that the hydrogen treatment of Hitco HMG-50 carbon fiber oxidized in nitric acid did not alter the shear strength although an increase in surface area accompanied the treatment. Thus, the parameter which is likely to be more significant in assessing the relative importance of resin-fiber interactions is the average number of bonds formed between the resin molecule and the surface groups per unit length of the fiber rather than the concentration of surface groups on the carbon fiber.

McKee and Mimeault [21] observed that the treatment increased the flexural strength of Modmor I carbon fiber by about 30% but decreased that of Modmor II by as much as 43%. The ILSS improved by a factor of 3 for acid-treated Modmor I and only slightly for acid-treated Modmore II carbon fiber (Figure 3). The best improvement is obtained when the time of treatment is very small. It is likely that the increased time of treatment causes excessive degradation of the fiber, as shown by Bahl et al. [53]. It is, however, not obvious from these results whether it is the increased surface area or the increased surface functionality which is responsible for improved shear properties. Therefore, Mimeault and McKee [27] outgassed oxidized carbon fibers by heat treatment in vacuum at 950°C (a treatment which almost completely removes the surface functional groups) [26] and studied shear properties of the composites with the evacuated carbon fibers. Little change in the shear properties was noticed, but there was a considerable increase in the surface area of the fibers (Table 6). It was, therefore, once again not possible to correlate the composite shear strength with either surface area or surface functionality because the increased surface area may compensate for the loss of surface functionality. This led McKee and Mimeault to argue that the improved shear properties may not be a direct result of the increasing interaction between the matrix and the carbon fiber but may be the result of removal of defects or flaws from the carbon fiber surface which act as stress enhancers and cause the failure of the composite.

Fitzer et al. [49], however, reported a direct relationship between the amount of carboxylic acid groups on the carbon fiber surface and the strength of the composite in the case of phenolic (as well as epoxy resin) resin as matrix and nitric acid oxidized carbon fiber. These workers observed that the oxidation of the carbon fiber improves its wetting by pure epoxy polymers and by pure hardener as well as with mixtures of both. This has been attributed to the chemisorption of the amine hardener on the carboxylic acid groups. The second amine group of the bifunctional hardener acts as a coupling agent by reacting with the epoxy group of the prepolymer. Thus the bonding between the cured matrix and the reinforcing fiber is enhanced, causing an improvement in ILSS of the resulting composite. This result has been confirmed by Ehrburger and Donnet [65] for oxidized fibers and epoxy composites.

These observations have also been supported by chemical analysis of hardener-treated carbon fibers after dissolving the excess of nonreacted hardener [130]. Kjehldahl analysis gave twice the concentration of amine groups compared with the concentration of carboxylic acid groups, indicating a stoichiometric chemical bonding between the diamine hardener and the carboxylic groups. Extraction of the hardener with formic acid showed the presence of hydrogen bonds. Thus, these workers believe [130] that the coupling mechanism involves the formation of hydrogen bonds either between the oxygen atom of the carboxylic group or between the hydrogen atom of the carboxylic group and the nitrogen atom of the amine group, and bonding of the prepolymer on the nitrogen atoms of both amine groups.

Several investigators [55,65,22,23] carried out detailed studies of the influence of various surface oxidations on the mechanical properties of carbon fiber reinforced polymer (CFRP) composites. They estimated the carbon fiber composition, the amounts of oxygen-containing functional groups, and surface areas of carbon fibers measuring the content of two types of surface groups (Table 7): (1) Strongly acidic groups such as carboxylic and phenolic which could be

Table 7 Composite Properties in Relation to Surface Acidic Groups on Carbon Fibers

	Surface group neutralized by			
Fiber	NaOH μEq/g	$NaOC_2H_5$ μEq/g	ILSS	Failure mode
AC nontreated	7	10	55	Shear
AC HNO_3	14	20	85	Intermediate
AC HNO_3 h.t.[a]	0	15	77	Intermediate
AC NaOH	16	16	92	Tensile
AC NaOH h.t.	0	13	77	Intermediate
AG nontreated	3	2	14	Shear
AG NaOH	16	17	50	Tensile
AG NaOH h.t.	2	21	51	Intermediate

[a]h.t., heat-treated.
AC Heat-treated at lower temperatures below 1500°C.
AG Heat-treated at high temperature 2000°C.
Source: Data from Ref. 22.

neutralized by sodium hydroxide, and (2) weakly acidic groups (hydroxylic) which were neutralized by sodium ethoxide. The presence of strong acidic groups determines the tensile failure mode of the composite, which is indicative of a strong interfacial bond. Introduction of weak acid groups enhanced the ILSS without a considerable alteration of the mode of failure. In other words, the nature of the interfacial bond remains almost unaltered.

In order to examine the effect of surface groups more closely, these workers degassed oxidized carbon fibers by heat treatment in argon at 1000°C (65,22]. This heat treatment removed the acidic groups almost completely. The weak acidic groups were, however, restored partly or completely on exposure of the degassed sample to air for one night at room temperature. ILSS studies of the composites prepared by using these carbon fibers showed a slight decrease, although the values still remained much higher than in the case of the untreated fiber (Table 7). This clearly indicates that the presence of acid groups on carbon fibers strongly influences the shear strength of the resulting composites. Clarke et al. [34] observed a relationship between the increase in ILSS of type I carbon fiber composites and the degree of oxidation in air, which they attributed partly to the increase in surface area and partly to the formation of acidic surface groups.

Bahl et al. [53] observed that the concentration of nitric acid and the time of treatment are important factors which determine the reinforcing properties of a carbon fiber. The oxidation involves three different processes, namely smoothing of the surface by removal of surface irregularities, formation of microetch pits, and the introduction of surface acidic groups. The smoothing of the surface increases the strength while the formation of etch pits tends to decrease it. Therefore, it is essential that the conditions of the treatment for obtaining optimum improvement in the strength of the composite be determined. The surface functional groups are polar in character and have definite structural alignments which may alter the interaction of the fiber-resin interface. Bahl et al. [53] observed that a 50-min time of treatment with 50% nitric acid were the optimum conditions for obtaining the best strength properties. If the oxidation is carried out for longer periods of time or with higher concentrations of nitric acid, the oxidation removes some of the more organized material from the carbon fiber surface, causing a decrease in strength. Fitzer and Weiss [131] also observed that the chemical bonding between the treated fiber and the matrix was decreased by secondary effect caused during extended oxidation treatment of the carbon fiber. Thus ILSS of the composite can decrease if the oxidative treatment is continued beyond the optimum conditions.

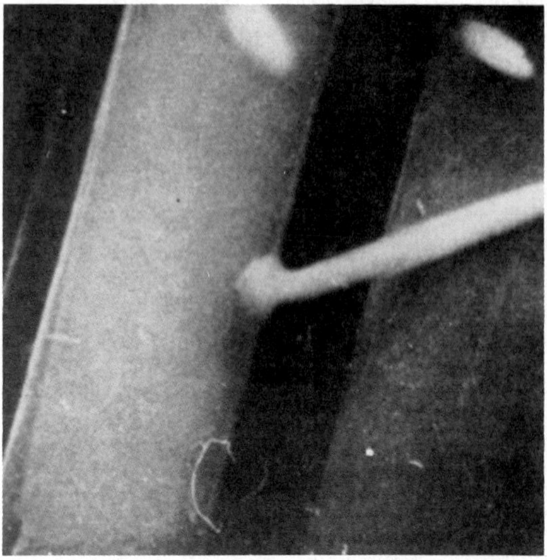

Figure 4 Single whiskers on carbon fiber ribbon (× 10,000). (From Ref. 66.)

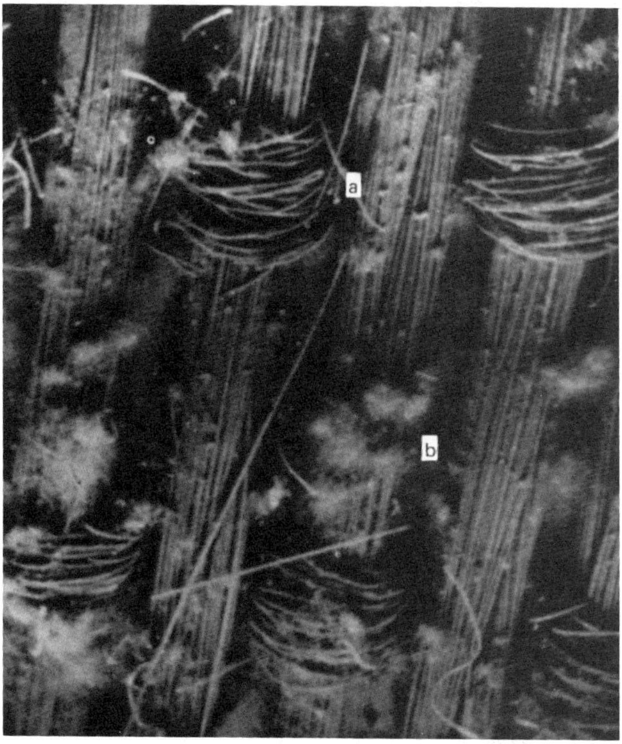

Figure 5 Hedgehogs (a) and balls (b) of whiskers on carbon fiber ribbon (a × 33; b × 700). (From Ref. 66.)

It appears that although the exact mechanism by which the oxidative treatments of carbon fibers improves the mechanical properties of the resulting composites is not fully elucidated and still a matter of speculation, a considerable amount of evidence has accumulated that the surface acidic functional groups, and the increase in specific surface [26,138] caused by the various oxidative treatments, improve bonding between the fiber and the resin matrix. This increases the wettability of the carbon fiber and enhances the interlaminar shear strength.

It is also evident that liquid-phase oxidation treatments are milder and do not cause drastic etch-pit formation, but are quite effective in creating surface functionality on the fiber surface. Consequently, these treatments should be further developed to give optimum conditions for improvement in composites properties.

3.4 NONOXIDATIVE SURFACE TREATMENTS

Several nonoxidative carbon fiber surface treatments have been developed and used with good results for improving the mechanical properties of the composites. These treatments include whiskerization, grafting of different kinds of polymers and certain forms of organic and polymer carbon coatings. However, a great deal of work on surface coatings is part of the patent and classified material.

3.4.1 Whiskerization

Whiskerization involves a nucleation process and the growth of very thin and high-strength single crystals of the chemical compounds, such as silicon carbide (SiC), titanium dioxide (TiO_2), and silicon nitride (Si_3N_4), on the fiber surface perpendicular to the fiber axis [19.66]. The whiskers grow from individual fibers and their growth on the fiber surface generally starts at the points of defects such as compositional heterogeneities, metallic inclusions or structural irregularities, and imperfections. Scanning electron microscopic examination [66] has shown that in the case of PAN carbon fibers the growth nucleus of the whisker is located in the valley or trough of the fiber. The growth of nuclei is a concurrent process and takes place preferably in one or two crystallographic directions. The whiskers may be single crystals (Figure 4) or aggregates in the form of hedgehogs or balls (Figure 5) which are formed by twisting of single crystals or by secondary nucleated crystals [66].

The whiskerization of a carbon fiber has been found to increase the interlaminar shear strength of the resulting CFRP composite significantly. A three- to fivefold increase in ILSS has been reported for rayon- and PAN-based carbon fibers [66–68]. However, in order to obtain maximum improvement, it is essential that the distribution of the whiskers be uniform over the whole surface. In addition, the mechanical properties of a whiskerized CFRP depend upon the amount of whiskers on the surface of the fiber, the macrostructure of the whiskers, and the carbon fiber surface. The ILSS was maximum when the whisker content was between 3 and 4% by volume although the shear modulus increased continuously with increase in the whisker content. The improvement in mechanical properties on whiskerization is attributed to a strong bonding of the whiskers to the graphite substrate and to the increase in interfacial area rather than to any increase in bonding between the fiber and the resin matrix. The decrease in ILSS with increased whiskerization is due to increased porosity of the CFRP. Although the treatment gives excellent results, it is not commercially used because it is expensive and difficult to carry out precisely.

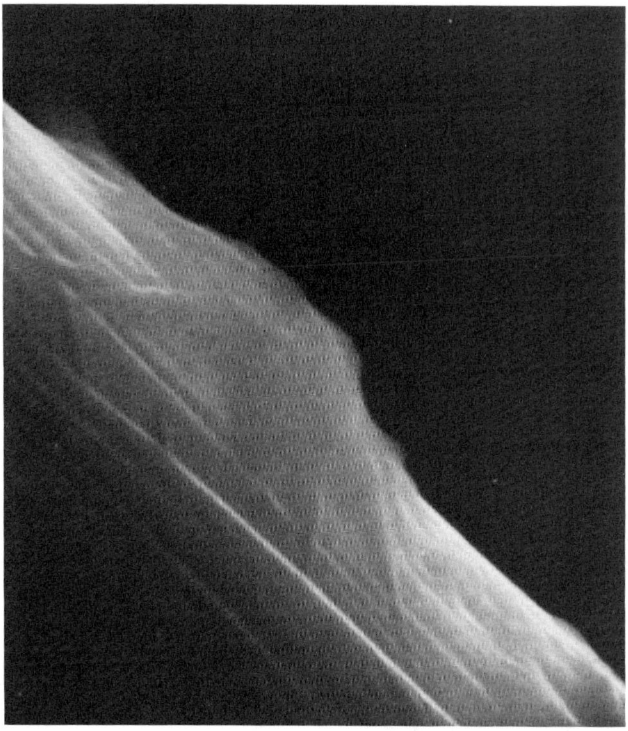

Figure 6 Scanning electron micrograph of PAN carbon fiber treated with a block polymer. (From Ref. 74.)

3.4.2 Polymer and Organic Surface Coating

Very important experimental work has been done to enhance the fiber-resin interfacial bonding by coating carbon fibers with organic polymeric materials such as polydivinyl benzene [69], polyamide [19,70], polyimide [71], organosilanes [19,72], and several alternating and block polymers [74]. The requirements for a polymer to be used as a coating are that it should be capable of being fixed on the carbon fiber surface by covalent, or by ionic bonds and be compatible with the resin which constitutes the matrix of the composite. To achieve better bonding between the resin and the polymer coating it is better if the polymer coating has functional groups capable of reacting with the resin.

Riess and co-workers [74] grafted alternating copolymers of styrene and maleic anhydride (SAM) and block polymers with SAM sequences on one end, which was compatible with the resin phase, and

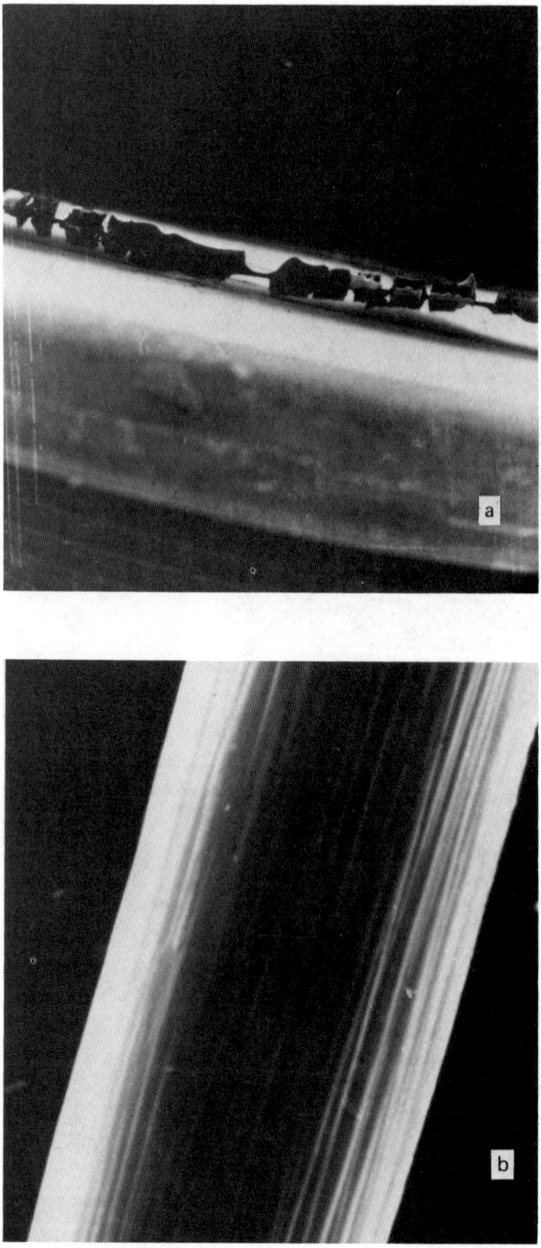

Figure 7 Scanning electron micrograph of electrocoated PAN-based carbon fiber (a) before washing and (b) after washing. (From Ref. 76, photo courtesy of R. V. Subramanian.)

an elastomer sequence on the other end, which dampened the stresses at the interface on the surface of type I and type II PAN carbon fibers. The carbon fibers were first oxidized in air or in ozone to introduce polar carboxylic groups on their surface, which were then neutralized by treatment with sodium hydroxide or sodium methoxide. The carbon fibers were then treated with a solution of SAM-alternating polymer, partially hydrolyzed and neutralized with sodium methoxide. In the case of block copolymers, the oxidized carbon fibers were treated with a dilute solution of the copolymer in benzene or in benzene-acetone mixture, where the copolymer exists essentially as a molecular dispersion, so that only a macromolecular layer of the polymer is obtained on the carbon fiber surface. The scanning electron micrograph (Figure 6) shows the fixation of a block copolymer as micelles on the carbon fiber surface. The composites obtained by using the treated carbon fibers showed a significant improvement in their shear strength and impact resistance. Although the exact mechanism of the process is not elucidated, the simplicity of the coating procedure makes it possible to treat long tows of carbon fibers continuously.

Subramanian et al. [75–77] investigated the effect of coating and grafting polymers on the carbon fiber surface during electro-initiated polymerization of a variety of monomers on their composite properties. The carbon fibers, passing continuously as anodes through an electrolytic cell, were coated with a number of polymers containing carboxyl functional groups. SEM micrographs of coated carbon fibers showed lumps of polymer of uneven thickness. However, when the carbon fibers were washed, the weight of the deposit was considerably reduced and reached a constant value after 5 minutes washing. The residual polymer coating was more tightly held and was more uniform (Figure 7). The coated carbon fibers were incorporated in an epoxy matrix. The composites showed a significant improvement both in interlaminar shear and impact strength (Table 8). The observed effects were controlled by molecular parameters such as the chemical composition, molecular weight, degree of cross-linking, and nature of the side chain in the polymer. The effect of polymer surface coatings on WyB-85 carbon fiber surface, preoxidized with 60% nitric acid, on the shear properties of the composites was also studied [21]. A marked improvement in shear properties was evident. Similar improvement in composite shear properties was obtained by coating oxidized Thornel carbon fibers with polyvinyl chloride and methane. But in this case 70% nitric acid was found to be more effective. However, no improvement in shear properties was noticed by coating carbon fibers with polyamide [70,19].

Cziollek and co-workers [78] modified type II PAN carbon fibers by coating the preoxidized fibers (oxidized with 65% nitric acid) with a liquid epoxy resin and a liquid diamine. The polypropylene matrix was also modified by creation of functional groups on the matrix. The

Table 8 Effect of Polymer Coating on Fortafil[a] Carbon Fiber-Epoxy Resin Composite Properties

Electrodeposited polymer[b]	Interlaminar shear strength (MPa)	Impact strength (MPa)	Flexural strength (MPa)
SMA 1000	68	57	110
SMA 2000	59	72	110
SMA 3000	62	56	100
PA 6	61	42	100
PA 18	52	44	91
AN 119	48	86	90
AN 169	54	140	86
Untreated	34	63	78
Surface treated	52	43	96

[a]Fortafil fiber (172 GN m^2 modulus), epoxy (DGEGA-MDPA) matrix.
[b]Copolymers of maleic anhydride with styrene (SMA), hexane (PA 6), octadecene (PA 18), and methylvinyl ether (AN).
Source: Data from Ref. 76.

composite shear strength improved between 30 and 50% when coated carbon fibers were used. But there was an additional improvement of about 60% when modified polypropylene matrix was used.

Harris et al. [73] treated preoxidized Modmor I carbon fibers (oxidized by boiling with nitric acid for 2 hr) with 30% by volume solutions of trialkoxy silane in methyl ethyl ketone solvent. The increase in weight of the carbon fiber after one hour contact with the solution was about 15%, which corresponded to a mean coating thickness of about 0.5 μm. The influence of silane coatings on the mechanical properties of the composites is shown in Table 9. It can be seen that a large portion of the improvement in shear properties can be attributed to the nitric acid treatment, the coating with silanes producing only a marginal improvement. Similar effect of chloro and alkoxy silane coatings was observed by Goan and Prosen [19] and Prosen and co-workers [72], who found that the shear strength showed only a slight increase in the case of preoxidized carbon fibers and a decrease in unoxidized carbon fibers.

3.4.3 Pyrolitic Surface Coatings

The effect of coating the carbon fiber surface by vapor-phase deposition of pyrolytic carbon, to enhance the interfacial adhesion with the resin matrix or to improve their oxidation resistance, has also been

Table 9 Effect of Surface Coating on the Mechanical Properties of Modmor

Fiber treatment ($V_f = 0.40$)	Tensile failure strength (kg/mm^2)	Interlaminar shear strength (kg/mm^2)	Ratio of shear strength to tensile strength
As-received	52.8	2.06	0.039
Boiled 2 hr in conc. HNO_3	56.6	2.52	0.045
Boiled in HNO_3 and coated with β-(3,4-epoxycyclohexyl-) ethyl-Si-$(OCH_3)_3$	63.7	2.77	0.044
Boiled in HNO_3 and coated with γ-glycidoxypropyl-Si-$(OCH_3)_3$	62.0	2.66	0.043
Boiled in HNO_3 and coated with mercaptopropyl-Si-$(OCH_3)_3$	39.5	1.82	0.046
Boiled in HNO_3 and coated with mercaptopropyl-Si-$(OCH_3)_3$	23.5	2.58	0.11

Source: Data from Ref. 21.

Table 10 Effect of Pyrolytic Carbon Coating of Carbon Fiber on Mechanical Properties of Composites

Carbon fiber	Flexural strength (10^6 psi)	Flexural modulus (10^3 psi)	ILSS (10^3 psi)
Modmor I uncoated	110	28	3–4
Mormor I coated	133	28.4	5.8
Modmor II uncoated	310 ± 10	19.0 ± 1	13.5 ± 5
Modmor II coated	310	19.0	12.2

Source: Data from Ref. 21.

studied by several workers [21,79–81]. McKee and Mimeault [21] coated high-modulus acrylic carbon fibers (Modmor I and Modmor II) with pyrolytic carbon by pyrolyzing acetylene or methane at 1000°C. The coating of Modmor I improved the composite shear properties but the treatment of Modmor II fibers decreased the shear strength (Table 10). Schmidt and Hawkins [79] also deposited pyrolytic carbon coatings on carbon fibers by passing the fiber through a mixture of methane and hydrogen at 1500°C. The hydrogen was added to control the rate of deposition and the thickness of the coating. However, an appreciable improvement in the composite shear properties was not obtained, although the oxidation resistance of the fiber improved considerably.

Marks et al. [80], coated Thornel 300 group carbon fibers (type I) with resinous coatings consisting of SC-1008 phenolic, Hercules 'H' polyaromatic, ferrocene in a phenolic resin binder (1:1), and Kerimid 601 polyimide. The first three materials were cured and pyrolyzed at 1000°C to yield carbonaceous and carbide coatings of varying morphology. The polyimide coating was cured but not pyrolyzed. The thickness of the surface coating varied with the material of the coating and was about 6 Å in the case of SC-1008 and between 6 and 15 Å for polyaromatic "H" resin. TEM analysis of the ferrocene phenol coating indicated a high degree of nonuniformity in texture. The coating was about 14 Å thick and exhibited a duplex structure with a variation from the interior to the exterior. SAD (selected area electron diffraction) showed that all the coatings had basal-plane orientation parallel to the fiber axis and appeared to have a lower density and a higher degree of porosity than the fiber. SEM examination showed no correlation between the surface morphology of the coating and the behavior of the composite. However, the composites obtained from coated fibers, particularly those obtained from pyrolyzed H polyaromatic resin and SC-1008 phenolic resin, showed significantly improved flexural strength retention compared to the untreated controlled fiber composite. Fitzer et al. [81] modified the surface of PAN carbon fibers, by coating with pyrocarbon depositions, to be used as prosthetics. The surface coating was carried out by decomposition of methane at 1300°C for a period of 3 minutes. The formation of carbon black was avoided by the addition of argon to methane. The average thickness of the coating, as measured by weighing, was 0.5 μm for Sigrafil HF with 10,000 monofilaments and 0.35 μm for Torray T300 with 3000 filaments. The coating increased the tensile strength of Torray by about 8% and of Sigrafil by about 4%. The Young's modulus and the strain to failure remained almost unaffected.

The structure of pyrolytic coatings deposited on cellulose, PAN, and pitch-based carbon fibers was studied by Gebhardt [82]. The surface coatings were formed by pumping a mixture of methane and hydrogen at one atmosphere and at low pressure (130 Pa) at a temperature of 1100°C for 80 hr through the system. In order to evaluate the influence of temperature and the nature of the carbon fiber surface on the structure of the coating, experiments were designed in which (1) the fibers were heated for one hour in argon at 3023 K prior to coating; (2) fibers coated at 1373 K were heated at 3023 K, and (3) preheated fibers were again heated after coating at 1373 K. The fragments of the coatings broken from the coated fibers were examined by thin film electron microscopy and SAD.

The surface coatings on cellulose-based T50 carbon fibers were isotropic but became oriented when the coatings on the preheated fibers (3023 K) were reheated. The coatings deposited at one atmospheric pressure were less adherent than those deposited at lower pressure. The coating conditions did not affect the structure of the coating. In the case of PAN carbon fibers, the coatings were both isotropic and anisotropic. Pitch fibers, however, formed anisotropic coatings under all conditions. The coatings on preheated pitch fibers and coatings heated after deposition were markedly anisotropic and crystalline. This indicates that the nature of the carbon fiber surface and its structure, rather than the deposition conditions, influence the structure of the coating. The deposition conditions only determined the thickness of the coating and the rate of its deposition.

Turner and Dietz [28] modified PAN carbon fibers by carbon vapor deposition at 1000°C in a flow system using acrylonitrile vapor with helium as the carrier gas. The surface was characterized by krypton adsorption at 77.4 K and found to be heterogeneous after the treatment. The deposition faithfully replicated the extent of surface of the original fibers but changed the fine grained porosity of the boundary surface.

Coating of carbon fibers with silica by the pyrolytic decomposition of tetraethyl orthosilicate at 900°C [18], and with elemental silicon by hydrogen reduction of trichlorosilane at 900°C has also been tried. These coatings did not improve the properties but were able to increase the oxidation resistance of the carbon fiber. A similar increase in oxidation resistance was observed when the carbon fibers are coated electrochemically with certain metallic coatings [18].

3.5 SURFACE FUNCTIONAL GROUPS ON CARBON FIBERS

The surface of a carbon fiber is an important region which determines its bonding behavior with the resin matrix. All surface treatments tend to modify this interfacial region by increasing the surface area, thus increasing the number of contact points, or by increasing the reactivity of the region by forming surface functional groups which provide a chemical type of bonding with the resin matrix. Although both factors, surface area and surface functionality, have been invoked to explain improvement in mechanical properties. It is difficult to assess quantitatively the relative importance of each parameter because both have been found to be enhanced by almost all the oxidative surface treatments. In more recent work, it has been indicated that there is a direct relationship between the number of acidic functional groups associated with the carbon fiber surface and the improvement in the shear strength of the fiber-resin composite. This has been attributed to the fact that the presence of these functional groups makes the carbon fiber surface more polar, thereby enhancing wetting and adhesion to the resin matrix. However, formulation of a definite mechanism for improvement in mechanical properties of the treated carbon fibers would require a precise knowledge of the composition of the carbon fiber surface, which includes the amount and the nature of the surface groups and their reactivities towards resin matrices. This can help future investigators to tailormake carbon fibers for a particular application by introducing varying amounts of the different types of surface groups.

The carbon fibers manufactured by the various manufacturers have a surface composition consisting of appreciable amounts of oxygen and traces of hydrogen and nitrogen. The surface composition is determined partly by the nature of the precursor fiber (rayon, PAN, or pitch) and partly by the reactivity of the fiber carbon with oxygen. The oxygen in carbon fibers may be derived either from the starting material and becomes part of the chemical structure of the carbon fiber as a result of imperfect carbonization, or becomes chemically bonded to the surface during activation or subsequent surface treatment. High-modulus carbon fibers, which are heat-treated at temperatures greater than 2000°C, contain only very minor quantities of of associated oxygen but pick up appreciable amounts during their surface treatments. These oxygen atoms are generally bonded to the carbon atoms at the edges and corners of the elementary graphitic crystallites or to carbon atoms at defect positions in the crystal lattice [133].

Although carbon-oxygen reactions and their thermodynamic and kinetic nature have been the subject matter of numerous investigations, definite information regarding the types of chemical oxygen containing functional groups, which are formed on carbon surfaces, is still not completely elucidated. Several hypothetical structures and formulae have been proposed for these carbon-oxygen surface compounds.

Many of these studies have used indirect methods of investigation involving acid-base neutralizations [83–87] and thermal decomposition of the surface compounds into CO_2 and CO. More recently however, emphasis has been placed on more direct analysis of the surface oxide layer by studying specific chemical reactions [88,89], as well as by employing infrared spectroscopy [90-92], electron spin resonance

(ESR) spectroscopy [93], internal reflection IR spectroscopy [94,95], and polarographic techniques [96,97]. These chemical and physicochemical methods of analysis have yielded a good deal of information regarding functional groups present on carbon surfaces. The more important groups that have been identified are carboxyl, phenol, lactone, carbonyl, and quinone. However, the different methods used for the determination of these groups have not yielded identical results and the presence of these groups could account for only a fraction of the total associated oxygen. Therefore, some of the workers [98] have questioned the application of organic chemistry to the reactions of surface compounds, while some others [99,100] are skeptical of the use of IR spectra in the case of carbons, as the wide absorption band can be interpreted in more than one way.

The investigations of surface oxygen groups on carbon fibers are comparatively recent, and much remains to be done so far. Hopfgarten [101] measured the atomic composition of external surfaces of type I, II, and III carbon fibers by Auger electron spectroscopy. The concentration gradients, determined by simultaneous ion etching of the fibers, indicated that the oxygen content of type I and type II fibers increased with depth up to 30 nm. However, the surface of type III carbon fibers contained more oxygen. Drzal [102,103] carried out mass spectral analysis of thermal desorption products on heat treatment of PAN carbon fibers before and after oxidation in air, and observed that the composition of the desorption products was different in each case. The major constituents were water and CO in untreated fibers and CO and CO_2 in the oxidized samples. The large amount of CO_2 in the oxidized samples was attributed to the decomposition of surface functional groups which were added to the surface during oxidation treatment.

Donnet and others [55,65,22,23,64] have carried out extensive work on the formation and estimation of oxygen functional groups on type I and type II acrylic carbon fibers under differing oxidation conditions. The treatment with nitric acid resulted in the chemisorption of appreciable amounts of oxygen in type II carbon fibers, but only small amounts in type I fibers. The oxygen was present in the form of acidic surface groups which were measured by titration with decinormal sodium hydroxide solution. ESCA studies indicated the presence of carboxylic and phenolic groups. The treatment of type I carbon fibers with Hummers' reagent (mixture of $NaNO_3$, $KMnO_4$, and H_2SO_4) created an appreciably larger amount of surface functionality (Table 11). It can be seen that even a moderate treatment of 3 minutes at 35°C increases the surface oxygen groups by about 70 times. As discussed earlier, graphitic oxide is formed by this treatment.

Elemental analysis and surface group determination of carbon fibers oxidized anodically in nitric acid and sodium hydroxide solutions [23] showed that the oxygen content increased appreciably only

Table 11 Development of Surface Functionality on Treatment of Type I PAN Carbon Fibers with Hummer's Reagent

Time of treatment t (min)	Temperature of treatment (°C)	Surface functionality (—COOH + —OH) μeq g^{-1}
0	35°	3
3	35°	200
3	40°	270
10	40	400
40	40°	835

Source: Data from Ref. 55.

in the case of the treatment in nitric acid. Analysis of the oxide layer carried out by differential neutralization technique showed the presence of two types of acidic groups with varying acidic strength [22]. Strongly acidic groups which titrated with sodium hydroxide were suggested to be carboxylic and phenolic, whereas weak acidic groups which titrated with sodium ethoxide were designated as hydroxylic and carbonylic. It has been observed that the oxidation in sodium hydroxide electrolyte produces strong acidic groups only while the treatment in nitric acid produces both types of surface groupings. The strongly acidic groups could be almost completely eliminated on heat treatment of the carbon fiber in argon at 1000°C, and did not reform on exposure of the degassed sample to air, but the weakly acid group could be reformed partially or completely on exposure to air.

Fitzer et al. [50] and Manocha [52] determined the composition of the surface oxide layer thermogravimetrically after oxidizing PAN (Sigrafil HM and HF) carbon fibers with 65% nitric acid and observed an increase in the surface oxide concentration with oxidation time. The total oxide concentration of the HM fiber treated for 36 hr was only one-fourth (1.5% by weight) compared to HF type after oxidation for 24 hr (6.5% by weight). The surface oxide decomposed to evolve CO_2 between 300 and 600°C and CO which started at 600°C in HF type and at 800°C in HM type fiber. The evolution of CO_2 has been attributed to the decomposition of carboxylic groups and the evolution of CO to carbonyl and hydroxyl groups at higher temperatures (600–1000°C).

Bahl et al. [53] also observed the formation of appreciable amounts of oxygen-containing surface groups on oxidation with nitric acid. The number of these groups increased with increase in time of treatment and concentration of the acid. The number and nature of the

Table 12 Amount of Burnoff During Oxidation Treatments and Amounts of Oxygen Evolved on Evacuating Untreated and Treated PAN Carbon Fibers at 1200°C

		O_2 evolved (g/100 g)				H_2 evolved (g/100 g)		
Fiber sample	Burnoff (%)	CO_2	CO	H_2O	Total	H_2O	H_2	Total
		Beslon Carbon Fiber						
Untreated	nil	3.80	0.30	0.18	4.18	0.02	2.22	2.24
Fiber treated with:								
60% Nitric Acid	9.1	7.60	0.40	1.80	9.80	0.20	1.25	1.45
70% Nitric Acid	27.5	6.80	0.60	0.98	8.38	0.11	0.95	1.06
Hydrogen peroxide	13.6	9.60	0.70	1.76	12.06	0.20	1.40	1.60
Potassium persulphate	6.8	4.90	0.22	3.40	8.52	0.38	nil	0.38
		Torray Carbon Fiber						
Untreated	nil	3.70	0.00	0.44	4.14	0.05	1.40	1.45
Fiber treated with:								
60% Nitric acid	8.1	6.50	0.30	1.41	8.21	0.17	1.08	1.25
70% Nitric acid	20.1	5.50	0.40	1.15	7.05	0.13	1.20	1.33
Hydrogen peroxide	9.1	7.80	0.40	1.95	10.15	0.22	1.11	1.33
Potassium persulphate	5.0	5.80	0.10	1.87	7.77	0.21	1.00	1.21
Ozonized oxygen	2.5	4.80	0.00	1.50	6.30	0.17	0.90	1.07

Source: Data from Ref. 58.

surface groups were determined by titration with sodium hydroxide. The surface groups decomposed to give CO_2 and CO on evacuation in nitrogen between 300 and 1000°C.

Bansal et al. [58] oxidized PAN-based carbon fibers with boiling nitric acid, hydrogen peroxide, potassium persulfate, and ozonized oxygen and observed that each oxidation treatment resulted in the chemisorption of appreciable amounts of oxygen which was evolved as CO_2, CO, and water vapor on evacuation (Table 12). Treatment with a mild oxidizing agent such as hydrogen peroxide results in the highest increase in oxygen content, while treatment with 70° nitric acid results only in appreciable burnoff of fiber material without showing any high chemisorption of oxygen. The efficiency of an oxidation treatment, in general, was found to depend on its capacity to cause burnoff of the carbon fiber.

Most of the chemisorbed oxygen was evolved as CO_2 on high-temperature evacuation and only a small part was evolved as CO and water vapor. The evolution of CO_2 commenced between 300 and 400°C and was completed at about 700°C whereas that of CO started at 600°C and was completed only at about 1000°C. This suggests that the oxygen in carbon is present in the form of different functional groups with varying thermal stabilities. The acidic character of the surface groups was determined by neutralization with sodium hydroxide solution. The surface groups were found to be partly acidic and partly neutral towards sodium hydroxide. The groups which are neutralized by sodium hydroxide are suggested to be carboxylic and phenolic, whereas the neutral groups may be quinonic and carbonylic. The carboxylic groups evolve CO_2 on evacuation, whereas the other groups evolve CO. Rand and Robinson [56,57] degassed PAN carbon fibers before and after oxidation with nitric acid and analyzed the evolved gases mass spectrometrically. The results indicated the evolution of appreciable amounts of CO_2 and CO, which were suggested to be the decomposition products of oxygen-containing functional groups. Similar evolution of CO_2 and CO was observed by Mimeault and McKee [27]. The presence of oxygen-containing functional groups such as carboxylic, phenolic, and carbonyl has also been indicated by chemical reactivity studies of rayon-based carbon fibers, before and after oxidation in air, with ionic solutes [20]. Fitzer [138] has made a systematic study of wet, dry, and anodic oxidative treatments using HM and HT carbon fibers and has identified reactive surface oxides by titration, desorption analysis, and selective blocking treatments.

3.6 ADSORPTION CHARACTERISTICS OF CARBON FIBERS

Protection of human beings against unfriendly chemical vapors is an important aspect of the adsorption process. Generally, activated carbon is incorporated into the protective garment. However, the methods are not entirely satisfactory. One potential way to solve the problem could be to incorporate activated carbon fibers in textile structures which could permit a far greater flexibility and simplification in the design of sorptive clothing. Thus a considerable amount of research effort is being directed towards studies relating to the adsorption characteristics of activated carbon fibers. Carbon fibers could also be potential molecular sieve materials, as they possess limited pore diameters.

Adams et al. [104] studied the adsorption of several hydrocarbon vapors such as benzene, n-hexane, cyclohexane, and dimethyl butane on carbon fibers prepared from Saran (a copolymer of PVC and PVDC) by carbonization at different temperatures. The fibers were essentially amorphous and nongraphitized. The adsorption was quite rapid in the case of benzene, hexane, and cyclohexane, very slow for dimethyl butane, but too slow for isooctane to obtain an adsorption isotherm. The mean pore volume of these fibers varied between 0.3 and 0.4 ml/g compared to 0.003 ml/g for rayon carbon fibers [9]. The lack of appreciable adsorption of isooctane indicates that these carbon fibers have pore diameters less than 10 Å.

Brooks and Scola [105] studied the chemical reactivity of several commercial rayon-based carbon fibers by the adsorption of water and several organic vapors, using chromatographic methods, before and after oxidation of the carbon fiber with nitric acid and air. The adsorption coefficients, calculated from the retention times for the various fibers, indicated that oxidation of the fibers enhanced the adsorption of polar adsorbates such as pyridine, water, aniline, and formic acid relative to hydrocarbons of comparable volatility. Heats of adsorption derived from the temperature dependence of the retention times did not vary appreciably with the treatment. The heats of adsorption of carbon tetrachloride, benzene, dioxane, and several other compounds by chromatographic methods were found to be similar to those found for graphitized carbon blacks [106]. Drzal [102,103] also measured the isosteric heats and the differential entropies of adsorption in PAN carbon fibers from krypton adsorption isotherms and showed an increase on oxidation of the carbon fiber with air. However, the increase was restricted only to a surface coverage of 30%.

Rand and Robinson [57] characterized PAN carbon fiber surfaces more precisely by studying the adsorption of n-butylamine, n-butyric acid, n-butanol, and 1-2 epoxy propane from n-heptane solutions using flow microcalorimetry. The adsorption of n-butanol and 1-2 epoxy propane was reversible and relatively unaffected by surface treatment of the fiber. However, the adsorption of n-butylamine was very strong, partially irreversible, and increased markedly by surface treatments. There was no adsorption of n-butyric acid. The area of exothermic peaks of irreversibly adsorbed n-butylamine, which was considered to be a semiquantitative measure of the surface acidity, increased by the manufacturer's surface treatment, but the external area of the carbon fiber was almost unchanged, which sug-

Table 13 Adsorption of Cationic and Anionic Solutes from Aqueous Solutions on HMG50 Carbon Fibers After Various Treatments

	Amount of cation adsorbed mol m^2			Amount of anion adsorbed mol M^2	
Fiber treatment	Li^+	Na^+	Methyl chloride	$PtCl^-_6$	Metanil yellow
Untreated	0.120	0.080	0.117	0.605	0.717
Oxidized in air	0.173	0.138	0.171	0.229	0.347
Oxidized in HNO_3 for 4 hr	—	0.599	0.738	0.260	0.266
Oxidized in HNO_3 for 8 hr	0.204	0.227	1.320	0.255	0.585
Oxidized in HNO_3 for 12 hr	2.10	1.380	1.090	0.117	0.420

Source: Data from Ref. 20.

gested that the fraction of the surface acidic groups had increased. HM fibers oxidized in nitric acid showed an increase in the adsorption of both reversibly and irreversibly adsorbed n-butylamine. However, the surface occupied by irreversibly adsorbed n-butylamine was greater than that occupied by the reversibly adsorbed one.

The adsorption of several cationic solutes, such as methylene blue chloride, sodium hydroxide, lithium hydroxide, and anionic solutes such as chloroplatinic acid and metanil yellow from aqueous solutions on rayon carbon fibers before and after oxidation with air and nitric acid, was investigated by Brooks et al. [20] using spectrophotometry and atomic absorption analysis. The oxidation with nitric acid or air resulted in a significant increase (up to 10 times) in the adsorption of cationic species but decreased the adsorption of anionic species by a factor of one sixth to one third (Table 13). There is, however, a large variation in the adsorption of different solutes which cannot be attributed to their molecular dimensions because methylene blue, the largest cation, shows the largest adsorption capacity for the 8 hr nitric acid treatment whereas lithium, the smallest ion, shows the largest adsorption capacity for the 12 hr oxidation. The adsorption of metanil yellow is favored over that of methylene blue in untreated and air oxidized carbon fibers, but this trend is reversed for the nitric acid-treated carbon fibers. The increase in the adsorption of cationic solutes has been attributed to the creation of acidic functional groups

Table 14 Surface Area, Iodine Adsorption, and Carbon Tetrachloride Capacity of Some Activated Carbon Fibers

Surface area (m^2 g)	Iodine adsorption capacity (mg g)	Carbon tetrachloride adsorption capacity (g/100 g)
425	371	9
692	694	—
1000	865	—
1245	1010	—
1365	1114	110
1600	1552	99.7
1685	1240	153.5
2000	1220	161

Source: Data from Ref. 107.

such as carboxyl and hydroxyl during the oxidation treatment. The decrease in the adsorption of anionic solutes on oxidation is attributed to the loss of surface carbonyl groups which provide hydrogen bonding to anionic solutes.

Economy and Lin [107] prepared activated carbon fibers from a highly cross-linked phenolic precursor fiber with BET surface area in the range 600–1200 m^2 g^{-1} and studied the adsorption of iodine, carbon tetrachloride, butane, phenol, and several gaseous species such as HCN, H_2S, and SO_2. The relationship between surface area, amount of iodine, and carbon tetrachloride adsorbed is shown in Table 14, suggesting the development of extremely fine micropores, which are not accessible to larger molecules of iodine or CCl_4. The adsorption isotherms of phenol from solution phase showed that the adsorption efficiency of activated carbon fibers was independent of its concentration, indicating that they could be particularly effective in removing trace contaminants from aqueous solutions. The adsorption of butane by carbon fibers was larger than its adsorption by granulated carbons. The activated carbon fibers were found to contain micropores in the range 7–16 Å radii in a large proportion, the transitional and macropores being almost absent. However, the spaces between fibers could act as macropores so that gas or the solution could pass through a bundle of fibers more easily than a granulated carbon bed. This also provided a more effective contact of the sorbate with the micropores and resulted in a faster rate of equilibration.

Adams and Boucher [25] activated Saran carbon fibers by heat treatment in CO_2 and air at 900°C and observed that activation increased the adsorption capacity of organic vapors such as benzene and isooctane at 25°C. Unactivated carbon fibers did not adsorb any isooctane, but appreciable amounts of isooctane could be adsorbed after activation. The amount of adsorption increased with increase in the burnoff. The increase in adsorption on activation was attributed to enlargement of the pores by elimination of decomposition products.

Aron et al. [108] and Macnair et al. [109] evaluated several commercial carbon fabrics as adsorbents for gases, vapors, and liquids by studying nitrogen adsorption, mercury porosimetry, and sorption of methylene blue from aqueous solutions. Pluton B-1 carbon fabric, which had a surface area of 217 m^2 g^{-1} and could adsorb about 17% of CCl_4 by weight, was found to contain mostly micropores with radii less than 10 Å. The carbon fabric appeared to be most suitable for gas and and vapor sorption and demonstrated a significant capacity for liquid-phase sorption.

3.7 WETTING CHARACTERISTICS OF CARBON FIBERS

The high-performance capability of a carbon fiber reinforced composite is certainly related to the wettability of its surface by the resin matrix. High-modulus carbon fibers are poorly wetted by epoxy resins and consequently result in low composite shear strengths. In order to improve wettability of these high-performance fibers, they are given several surface treatments to introduce polar surface groups which produce a firmer bond between the fiber and the matrix.

A precise measurement of wettability of fibrous material is a difficult process although several procedures have been proposed. These methods include several static methods which directly measure the contact angle between the liquid and the filament [110,111]. Grindstaff [110] developed a simple apparatus which involved a simple technique based on the level surface method and could obtain either advancing or receding contact angles. However, these contact angle measurements are not accurate because the carbon fiber filaments are extremely fine. The use of a single filament involves several handling and contamination problems.

Dyba et al. [112] and Kemmel and Steiger [113], described dynamic methods which consist of passing a single filament through a liquid bath under such conditions that there is no distortion of the liquid interface around the filament. These methods also involved single filaments.

Bobka and Lowell [16] determined the wettability of carbon fibers by measuring the wicking rates either by the mass pick-up technique or by the surface velocity method. The contact angle was then calculated using the Washburn equation [114] derived for a system of uniform capillaries.

$$\frac{dh}{dt} = \frac{r\gamma \cos \theta}{4\eta h} - \frac{r^2 g\rho}{8\eta}$$

where dh/dt is the wicking rate in a capillary of radius 2, and ρ, γ, and η are the liquid density, surface tension, and viscosity, respectively, and g is the acceleration due to gravity.

Chwastiak [115], however, felt that Washburn's simple model is qualitative for wicking in yarn because the numerical value of the "equivalent diameter" of a capillary in the bed cannot be determined. He modified the procedure for wicking rate measurements by enclosing the carbon fiber bundle in a glass tube so that the porosity is fixed for a given strand of carbon fibers. The fiber is loaded into the tube by looping a piece of a string around the middle of the strand of about 50 fibers, threading the string ends through the tube, and pulling the fiber strand into the tube so that about 100 lengths of the fiber are loaded into the tube. The excess yarn is cut away to leave about 2 mm of the fiber yarn protruding from the other end. The loaded yarn is hung from the arm of an analytical balance with the protruding ends pointing downwards. A pan containing the liquid is raised up so that the protruding ends of the caron fiber are just dipping into the liquid. The liquid wicks up into the fibers. The weight gain is determined as a function of time until a constant weight is obtained. The weight change is corrected for buoyancy and surface tension acting on the ends of the yarn by determining the weight of the yarn in air at the end of the experiment. The difference between the weight of the bundle with ends immersed in the liquid and with ends taken out of the liquid is the amount of correction.

The wicking of a liquid in a fiber bundle is caused by surface forces and is resisted by mechanical forces which are viscous, gravitational and inertial in origin. The decrease in the interfacial free energy when a liquid moves a distance dh within the yarn bundle is balanced by the mechanical work required to move the liquid through that distance so that the free energy of wetting $\Delta\gamma$, can be written as

$$\Delta\gamma = \frac{1}{C_f}\left(\frac{dw_v}{dh} + \frac{dw_g}{dh} + \frac{dw_i}{dh}\right)$$

$$= \frac{1}{C_f}(\Phi_v + \Phi_g + \Phi_i)$$

where dw_v, dw_g, dw_i are viscous, gravitational, and inertial work, respectively, and Φ_v, Φ_g, and Φ_i are the respective corresponding forces. The expressions for these forces were derived by Chwastiak, and if we neglect the gravitational and inertial forces, which are small, in such wicking rate measurements, the expression for the free energy of wetting could be written as

$$\Delta\gamma = \frac{128(1-\varepsilon)^2 h^2 \rho_f \eta \alpha}{K^2 \varepsilon^2 v W_f D_f \rho^2} \cdot \left(\frac{m^2}{t}\right)$$

where W_f and D_f are the weight and diameter, respectively, of the fiber bundle; ρ_f and ρ are respectively densities of the fiber and the wetting liquid; η is the viscosity of the liquid, v is the volume, and h is the height of the fiber bundle, ε is the porosity of the bundle, α is the slope of wicking rate data m versus t on log-log plot where m is the mass of the liquid wicked in time t. K is a hydraulic constant which relates the average equivalent diameter of a pore in a fiber bundle to the volume and the surface area of the pore.

The parameters regarding the geometry of the fiber bundle and the properties of the liquid are readily obtained. The hydraulic constant K was determined by Chwastiak [115] to be 3.3. The free energy of wetting was converted into contact angles by using

$$\cos\,\theta = \frac{\Delta\gamma}{\gamma}$$

where γ is the surface tension of the wicking liquid. These wicking rate calculations were used successfully by Chwastiak [115] to evaluate wettability of Thornel 50 carbon fiber by water, glycol, heptane, and epoxy resins. The measurements were limited to systems where the contact angle was less than 90°. The reproducibility was within 7% between tests.

Schultz et al. [116] and Cazeneuve [117] have recently developed a tensiometric method for determining the wettability of low-diameter carbon fibers. The method involves two immiscible liquid phases which make it possible to measure finite angles of contact. It measures the variation of forces both in the static and dynamic regime when a solid is immersed and taken out through the liquid-liquid interface. The measurement in static regime determines the various components of the surface energy, whereas the measurement in dynamic regime could provide information on the state of heterogeneity of the fiber surface. Two samples of carbon fibers T-300 and M-40 were examined using formamide associated with a series of nonpolar liquids.

Fitzer et al. [132] measured the advancing and receding contact angle between carbon fibers and several organic liquids using a gravimetric method. The monofilament was brought into contact with the wetting liquid by raising a liquid container and pulled away by lowering the liquid container. The adhesive and pull-out forces F_1 and F_2, respectively, were measured and contact angles measured by using

Table 15 Contact Angles of Carbon Fibers in Polystyrene Before and After Surface Treatment in Relation to ILSS in Polystyrene-Based Composites

Carbon fiber	Treatment	Contact angle	ILSS
Sigrafil HF HTT 1600°C	Untreated	0	17
Sigrafil HM HTT 2800°C	Untreated	15	13
Sigrafil HM	HNO_3 treated	0	14
Modmor II	Untreated	0	18
Modmor II	SiC coated	34	20
Modmor II	SiC + Silan A-174	0	24

Source: Data from Ref. 118.

$$\cos\theta = \frac{F_1}{C_f \cdot \gamma} = \frac{F_2}{C_f \cdot \gamma}$$

where C_f is the circumference of the filament and γ is the surface tension of the liquid. Both type I and type II carbon fibers were completely wetted by acetone, chlorobenzene, and monomers of resin. However, a difference in the wetting behavior was observed for polystyrene before and after treatment of the carbon fibers. Graphitization decreased the wettability (increased the contact angle) while treatment with nitric acid increased it (Table 15). Coating with silicone carbide decreased wettability and the treatment with a silicon coupling agent increased the wettability. The interlaminar shear strength of the composites increased with each treatment. The increase in ILSS in the case of nitric acid is in line with increased wettability, but the increase in case of silicon carbide treatment, in which case the wettability decreases, has been attributed to increase in roughness of the surface. The oxidation with nitric acid also increased the wettability of type I and type II carbon fibers towards epoxy resin, hardener, and a mixture of the resin and the hardener [49]. This improvement has been attributed to the chemisorption of the amine group of the hardener or the epoxy resin on the carboxylic acid groups created by the oxidation treatment.

Bansal and Chhabra [119] have recently calculated contact angles between PAN carbon fibers and several liquids including epoxy resin (araldite) from wicking rate measurements using Chwastiak equation and observed that the oxidation results in enhancing the wettability, decreasing the contact angle from 79° in untreated fiber to 47° in oxidized fiber. Similar decreases in contact angle in Thermolon carbon fibers have been observed by Yamamoto et al. [120], who measured contact angle between the carbon fiber and araldite. Yamamoto [139] has recently shown a relationship between ILSS of a composite and wettability of the matrix.

3.8 INTERCALATION OF CARBON FIBERS

Carbon fibers, although generally nongraphitic and nongraphitizing, form intercalation compounds with alkali metals such as potassium and cesium, nitric acid, and halogens. Herinckx et al [121–123] studied the intercalation of sodium, potassium, and cesium vapors into rayon-based carbon fibers, WyB and WyD (Thornel) heat-treated to 2800°C, using a thermogravimetric balance at several temperatures between 125 and 300°C. The structure of the intercalated carbon fiber was examined by x-ray diffraction fiber diagrams. Sodium is taken up and retained by the WyD carbon fibers in appreciable amounts (ratio C/Na = 61 at 300°C and 99 at 400°C) but apparently does not form interstitial compounds. However, potassium forms well-defined intercalation compounds (Figure 8) of the first stage with composition C_8K (insertion between every pair of carbon layers) and of the second stage with composition $C_{16}K$ to $C_{24}K$ (insertion between every second pair of carbon layers). The color of the carbon fiber changed from

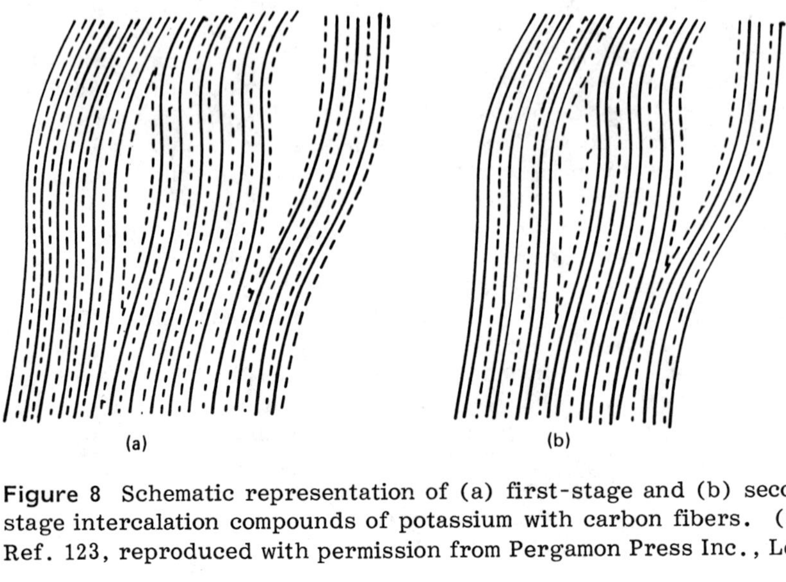

Figure 8 Schematic representation of (a) first-stage and (b) second-stage intercalation compounds of potassium with carbon fibers. (From Ref. 123, reproduced with permission from Pergamon Press Inc., London.)

Table 16 Residual Intercalated Halogen in Various Carbon Fibers

Carbon fiber	Precursor material	Residual halogen mol/1000 mol carbon	
		Br_2	ICl
F 46 HTT 1800°C	Pitch	0	15
Celanese 1800°C	PAN	0	18
F 54 HTT 3000°C	Pitch	15	15
F 54 II HTT 3000°C	Pitch	15	15

Source: Data from Ref. 124.

black to steel blue for the second stage and to bronze brown for the first-stage intercalation compounds. X-ray studies showed potassium layers in both first and second-stage compounds to be hexagonally close packed. In the case of second-stage compounds the change in the composition from $C_{16}K$ to $C_{24}K$ is caused by an increase in the concentration of randomly distributed vacancies in the two-dimensional potassium lattice with progressive desorption.

The reaction with potassium is slow below 200°C and the equilibrium value was not obtained in a reasonable time. However, above 250°C the reaction attained equilibrium in a few minutes. At temperatures above 200°C and potassium pressure greater than 10^{-3} torr, the carbon fibers retained appreciable amounts of potassium above the equilibrium value. The excess of potassium (greater than the composition C_8K) increased with increasing temperature and was higher for WyB than for WyD carbon fiber. The maximum of the excess corresponded roughly to the monoatomic layer of potassium on the internal surface due to the micropores in the carbon fibers. This internal surface was larger for WyB than for WyD. Thus this excess of potassium was considered to be adsorbed at the internal surface involving lower energies of interaction than those of the intercalated potassium atoms. The intercalation with potassium did not alter the essential characteristics of the preferred orientation, the layer size, the layer stacking, and the pore structure. The mechanical properties of the carbon fibers were also not affected significantly although electrical conductivity was appreciably increased by intercalation of potassium. The intercalation of cesium followed the same behavior as that of potassium.

Hooley and Dietz [124,125] recently studied the intercalation of bromine and ICl in graphitized PAN and pitch carbon fibers, heat-treated at different temperatures between 1800 and 3000°C, by exposing the carbon fibers to saturated bromine or ICl vapors or to a 3M solution of bromine in nitromethane. The treated samples were then

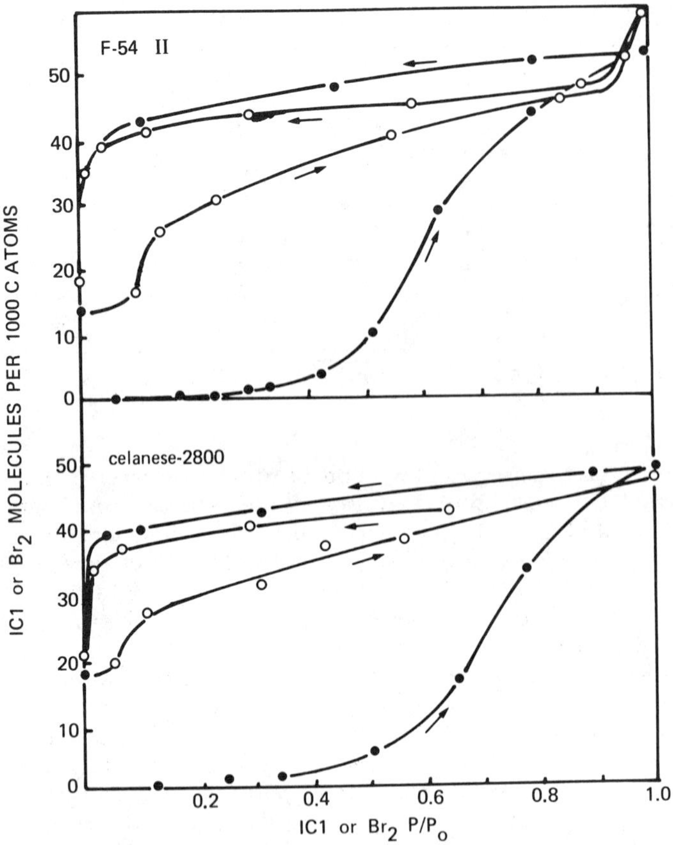

Figure 9 Intercalation of ICl followed by bromine on carbon fiber. (From Ref. 124, reproduced with permission from Pergamon Press Inc., London.)

exposed to vacuum or air at 20°C or higher when most but not all of the halogen had been removed from the structure to obtain a carbon fiber with residual intercalated bromine or ICl. The amounts of the residual halogens for different carbon fibers are shown in Table 16. It is seen that Celanese 2800°C and F-46 carbon fibers heat-treated at 1800°C do not intercalate bromine presumably because their structures were not sufficiently ordered or because of compressive stress in the surface region. However, these carbon fibers intercalate ICl, which has a lower threshold pressure than bromine for intercalation in well-ordered graphite.

The intercalation isotherms of Br_2 and ICl in graphitized carbon fibers showed that the threshold pressure on the intercalated fiber is lower than on the original fiber. This is attributed to the formation of microscopic cracks in the interlocking amorphous carbon networks between the graphitized domains and along the length of the carbon fibers during first intercalation. Therefore, when the intercalated carbon fiber is again exposed to bromine, the compression tension build up is lower and hence the rate of intercalation is higher. In the case of 2800°C Celanese PAN fiber, which has a lower order than the 3000°C fiber, the initial threshold pressure will be higher and consequently it may not intercalate bromine. However, when this fiber (2800°C Celanese) is first intercalated with ICl, which has a lower threshold pressure, the intercalation produces cracks so that bromine can now intercalate the structure during the second run (Figure 9).

X-ray diffraction measurements of the interplanar layer spacing for two carbon fibers before and after intercalation with bromine showed that the spacing was greater by 0.002 nm in the intercalated fibers.

Donnet and co-workers [22,23] and Raedersdort [126] have observed that oxygen intercalates into acrylic type I fiber when the oxidation is carried out electrochemically using an acidic or alkaline medium. Vogel [127,129] and Vogel and co-workers [128] observed that high-modulus carbon fibers can be readily intercalated by immersion in red fuming nitric acid. The measurements of conductivity of the carbon fibers before and after intercalation showed that the process is very fast and takes place in only a few minutes at room temperature. No intercalation was observed in the case of pitch-based carbon fibers with poor orientation of the graphitic crystallites.

REFERENCES

1. Perret, R., and Ruland, W. J. Appl. Cryst. *3*, 525 (1970).
2. Fourdeux, A., Perret, R., and Ruland, W. *Carbon Fibers—Their Composites and Applications*. The Plastics Institute, London, 1971, p. 57.
3. Perret, R., and Ruland, W. J. Appl. Cryst. *2*, 209 (1969).
4. Johnson, D. J., and Tyson, C. N. Brit. J. Appl. Phys. *D3*, 526 (1969).
5. Johnson, D. J. *Carbon Fibers—Their Composites and Applications*. The Plastics Institute, London 1971, p. 52.
6. Badami, D. V., Joiner, J. C., and Jones, G. A. Nature *215*, 386 (1967).
7. Donnet, J. B., and Ehrburger, P. Carbon *15*, 143 (1973).
8. Bacon, R., and Silvaggi Carbon *9*, 321 (1971).
9. Didchenko, R. *Carbon and Graphite Surface Properties Relevant to Fibre Reinforced Composites*. AFML-TR-68-45, Air Force Materials Laboratory, (February 1968).

10. Spencer, D. H. T., Hooker, M. A., Thomas, A. C., and Napier, B. A. *Proc. 3rd London Conf. on Ind. Carbon and Graphite*, Soc. Chem. Ind., London, 1971, p. 467.
11. Christner, L. G., and Walker, P. L., Jr. *10th Bienn. Conf. on Carbon*, Lehigh University, Bethlehem, Pa, 1971, summary of papers, p. 43.
12. Johnson, J. W., Rose, P. G., and Scott, G. *Proc. 3rd London Conf. on Ind. Carbon and Graphite*, Soc, Chem. Ind., London, 1971, p. 443.
13. Adams, L. B., Boucher, E. A., Cooper, R. N., and Everett, D. H. *Proc. 3rd London Conf. on Ind. Carbon and Graphite*, Soc. Chem. Ind., London, 1971, p. 478.
14. Molleyre, F., and Bastick, M. *Proc. 4th London Intern. Conf. on Carbon and Graphite*, Soc. Chem. Ind., London, 1974, p. 190.
15. Johnson, A., and Bradshaw, W. G. *13th Bienn. Conf. on Carbon*, Irvine, Ca., 1977, extended abstracts, p. 98.
16. Bobka, R. J., and Lowell, L. P. *Integrated Research on Carbon Composite Materials*, AFML-TR-66-310 Part I, Air Force Materials Laboratory, (October 1960), pp. 145—152.
17. Pallozzi, A. A. Soc. Plastics Engg. J. *8*, February 1966.
18. Herrick, J. W., Grüber, P. E., Jr., and Mansur, F. T. *Surface Treatments for Fibrous Carbon Reinforcements*, AFML-TR-66-178 Part I, Air Force Materials Laboratory (July 1966).
19. Goan, J. C., and Prosen, S. P. In *Interface in Composites*, ASTM Special Tech. Publ. No. 452, p. 3, (quoted Ref. 57 McKee Review).
20. Brooks, C. S., Golden, G. S., and Scola, D. A. Carbon *12*, 609 (1974).
21. McKee, D. W., and Mimeault, V. J. *Chemistry and Physics of Carbon*, vol. 88 (Edited by P. L. Walker, Jr. and P. Thrower), Marcel Dekker, New York, 1973.
22. Ehrburger, P., Herque, J. J., and Donnet, J. B. *5th London Intern. Carbon and Graphite Conf.*, vol. 1, Soc. Chem. Ind., London, 1978.
23. Ehrburger, P., Herque, J. J. and Donnet, J. B. *Petroleum Derived Carbons* (Edited by M. L. Deviney and T. M. O'Grady) ACS Symposium Series No. 21, 1975, p. 324.
24. Perret, R., and Ruland, W. *9th Bienn. Conf. on Carbon*, Boston, Mass., 1969, summary of papers SS-22.
25. Adams, L. B., and Boucher, E. A. Carbon *16*, 75 (1978).
26. Bansal, R. C., and Chhabra, P. Indian J. Chem., *20A*, 449 (1981).
27. Mimeault, V. J., and McKee, D. W. Nature *224*, 793 (1969).
28. Turner, N. H., and Dietz, V. R. Carbon *11*, 256 (1973).
29. Kaelble, D. H., Dynes, P. J., Crane, L. W., and Maus, L. Adhesion 7, 25 (1975).

30. Larsen, J. V., Smith, T. G. and Erickson, P. W. Carbon fiber surface treatment, NOLTR 71-165 (1971).
31. Daukeys, R. J. J. Adhesion 5, 211 (1973).
32. Eckstein, B. H. *13th Bienn. Conf. on Carbon*, Irvine, Ca., 1977, summary of papers, p. 433.
33. Novak, R. C. In *Composite Materials: Testing and Design* ASTM Special Tech. Publ. No. 460, 1969, p. 540.
34. Clark, D., Wadsworth, N. J., and Watt, W. *Carbon Fibers—Their Place in Modern Technology*, The Plastics Institute, London, 1974, p. 44.
35. Sach, R. S., and Bromley, J. U. S. Patent 3,642,513 (February 1972).
36. Scola, D. A., and Basche, M. U. S. Patent 3,720,536 (March 1973).
37. Druin, M. L., Ferment, G. R., and Rao, V. N. P. U.S. Patent 3,754,957 (August 1973).
38. Cunningham, A. L. U.S. Patent 3,816,598 (June 1974).
39. Harker, H., Marsh, H., and Wynne-Jones, W. F. K. *Proc. First Conf. on Ind. Carbon and Graphite* Soc. Chem. Inc., London, 1958, p. 291.
40. Harker, H. *Proc. 4th Carbon Conf.*, Pergamon Press, New York, 1960, p. 125.
41. Walker, P. L., Jr., Shelef, M., and Anderson, R. A. *Chemistry and Physics of Carbon*, vol. 4 (Edited by P. L. Walker, Jr.) Marcel Dekker, New York, 1968, p. 287.
42. Kalnin, I. U.S. Patent 3,723,607 (March 1973).
43. Bansal, R. C., and Chhabra, P. Unpublished data.
44. Bansal, R. C., and Chhabra, P. Unpublished work.
45. McKee, D. W. Carbon *8*, 131 (1970).
46. McKee, D. W. Carbon *8*, 623 (1970).
47. Herrick, J. W. 23rd Annual Tech. Conf. SPI Reinforced Plastics/Composites Division, Section 16A, February 1968.
48. Scola, D. A., and Brooks, C. S. 15th Soc. Plastics Ind. Meeting, Washington, D.C., February 1970.
49. Fitzer, E., Geigel, K. H., Hüttner, W., and Weiss, R. *14th Bienn Conf. on Carbon*, Pennsylvania State University, University Park, Penn. 1979, extended abstracts, p. 220.
50. Fitzer, E., Geigel, K. H., and Hüttner, W. *14th Bienn. Conf. on Carbon*, Pennsylvania State University, University Park, Penn., 1979, summary of papers, p. 236.
51. J. Cziollek, J., Fitzer, E., and Weiss, R. *Carbon '80 3rd Intern. Carbon Conf.*, Baden-Baden, 1980, Preprints p. 632.
51a. Cziollek, J., Fitzer, E., and Weiss, R. Intern. Symp. on Carbon, Toyohashi, Japan 1982.
52. Manocha, L. M. *Carbon '80 3rd Intern. Carbon Conf.*, Baden-Baden, 1980, preprints of papers, p. 633.

53. Bahl, O. P., Mathur, R. B., Seth, R. L., and Dhami, T. L. *Carbon '80 3rd Intern. Carbon Conf.*, Baden-Baden, 1980, preprints of papers, p. 602.
54. Donnet, J. B. R. G. C. P. *50*, 51 (1973).
55. Donnet, J. B., Papirer, E. and Dauksch, H. *Carbon Fibres—Their Place in Modern Technology*, The Plastics Institute, London, 1974, p. 58.
56. Rand, B., and Robinson R. Carbon *15*, 257 (1977).
57. Rand, B., and Robinson, R. Carbon *15*, 311 (1977).
58. Bansal, R. C., Chhabra, P., and Puri, B. R. Indian J. Chem. *19A*, 1149 (1980).
59. Well, H., and Colclough, W. F. U.S. Patent 3,657,082 (April 1972).
60. Druin, M. L., and Dix, R. U.S. Patent 3,894,884 (July 1975).
61. Joo, L. A., Prescott, R., Sharper, G. E., and Snodgrass, F. E. U.S. Patent 3,989,802 (November 1976).
62. Rieux, J. P., and Lehurreau, J. Fr. Patent 72.14.950 (1972).
63. Courtaulds Ltd. Br. Patent 14.4341 (1964).
64. Ehrburger, P., Herque, J. J., and Donnet, J. B. *Proc. 4th London Intern. Carbon and Graphite Conf. 1974*, Soc. Chem. Ind., London, 1976, p. 201.
65. Ehrburger, P., and Donnet, J. B. Phil. Trans. R. Soc. London *A294*, 495 (1980).
65a. Yamamoto, M. Intern. Symp. on Carbon, Toyahashi, Japan, 1982.
66. Rabotnov, J. N., Perov, B. V., Lutsan, V. G., Ssorina, T. G., and Stepanitsov, E. I. *Carbon Fibres— Their Place in Modern Technology*, The Plastics Institute, London, 1974, p. 65.
67. Shaver, R. G. A. I. Ch. E. Materials Conf. Philadelphia, April 1968. (Quoted in McKee & Mimeault, review, Ref. 72).
68. Simon, R. A. , and Bosen, S. P. 23rd Ann. Tech. Conf. Soc. Plastics Ind. Reinforced Plastics Division Section 16B, February 1968 (quoted in McKee Ref. 74).
69. Trostyanskaya, E. B., and Kobets, L. P. Plast. Massy *1*, 53 (1970).
70. Herrick, J. W. 12 Natl. Symp. Soc. Aerosp. Mater. Process Eng. Anaheim, Calif. October 1967, paper AC-8.
71. Marks, B. S., Mauri, R. E., and Bradshaw, W. G. *12th Bienn. Conf. on Carbon*, Pittsburgh, Penn., 1975 extended abstracts, p. 337.
72. Prosen, S. P., Duffy, J. V., Erickson, P. W., and Kinua, M. A. 21st Soc. Plastics Ind. Ann. Tech. Conf. 1966 Section 8-D, (quoted in McKee review Ref. 60).
73. Harris, B., Beaumont, P. W. R., and Prosen, A. J. Mater. Sci *4*, 432 (1969).

74. Riess, G., Bourdeux, M., Brie, M., and Jouquet, G. *Carbon Fibres—Their Place in Modern Technology*, The Plastics Institute, London, 1974, p. 52.
75. Subramanian, R. V. *Carbon '80 3rd Intern. Carbon Conf.*, Baden-Baden, 1980, preprints, p. 690.
76. Subramanian, R. V., Sundaran, V., and Patel, A. K. *33rd Annual Tech. Conf. 1978 Reinforced Plastics/Composites Instt.* Soc. Plastics Ind. Inc. Section 20, p. 1–8.
77. Subramanian, R. V., and Jakubowski, J. J. Polymer Engg. and Sci. *18*, 590 (1978).
78. Cziollek, J., Fitzer, E., and Weiss, R. *Carbon '80 3rd Intern. Carbon Conf.* Baden-Baden 1980. Preprints p. 632.
79. Schmidt, D. L., and Hawkins, H. T. Filamentous carbon and graphite, AFML-TR-65-160, Airforce Materials Laboratory, August 1965.
80. Marks, B. S., Mauri, R. E., and Bradshaw, W. G. *12th Bienn. Conf. on Carbon*, Pittsburgh, Penn., 1975 extended abstracts p. 337.
81. Fitzer, E., Hüttner, W., and Wolter, D. *13th Bienn. Conf. on Carbon*, Irvine, Calif., 1977, extended abstracts, p. 180.
82. Gebhardt, J. J. *14th Bienn. Conf. on Carbon*, Pennsylvania State University, University Park, Penn., 1979, extended abstracts, p. 232.
83. Puri, B. R., and Bansal, R. C. Carbon *1*, 451 (1964).
84. Puri, B. R., and Bansal, R. C. Carbon *1*, 457 (1964).
85. Barton, S. S., Gillespie, D., and Harrison, B. H. Carbon *11*, 649 (1973).
86. Donnet, J. B., Lahaye, J., and Benardin, J. Bull. Soc. Chim. France *5*, 1790 (1969).
87. Boehm, H. P. Adv. Catalysis *16*, 179 (1966).
88. Puri, B. R. *Chemistry and Physics of Carbon*, vol. 6, (Edited by P. L. Walker, Jr.) Marcel Dekker, New York.
89. Donnet, J. B., and Voet *Carbon Black*, Marcel Dekker, New York.
90. Donnet, J. B., Hueber, H., Perol, N., and Jaeger, J. J. Chim. Phys. *60*, 426 (1963).
91. Smith, R. N., Young, D. A., and Smith, R.A. Trans. Faraday Soc. *62*, 2280 (1966).
92. Cole, H. M., and Dannenberg, E. M. Rubber Age *81*, 1502 (1958).
93. Harker, H., Gallagher, J. T., and Parkin, A. Carbon *4*, 401 (1966).
94. Mattson, J. S., Mark, H. B., Jr., and Weber, W. J. Anal. Chem. *41*, 355 (1969).
95. Mattson, J. S., and Mark, H. B., Jr. J. Colloid and Interface Sci *31*, 131 (1969).

96. Donnet, J. B., and Henrich, G. Bull. Soc. Chim. France 1609 (1960).
97. Hallum, J. V., and Drushel, H. B. J. Phys. Chem. *62*, 110 (1958).
98. Boehm, H. P., Diehl, E., and Heck, W. Symposium on Carbon, Tokyo, Japan VIII-10-1 (1964).
99. Kiselev, A. V. *Physico-Chimie du Noir de Carbone*, Mulhouse, France, 1963, p. 15.
100. Harker, H. *Proc. 4th Carbon Conf.*, Pergamon Press, London, 1960, p. 125.
101. Hopfgarten, F. 13th Bienn. Conf. on Carbon, Irvine, Calif., 1977, extended abstracts, p. 288.
102. Drzal, L. T. Carbon *15*, 129 (1977).
103. Drzal, L. T. Carbon *17*, 375 (1979).
104. Adams, L. B., Boucher, E. A., and Everett, D. H. Carbon *16*, 75 (1978).
105. Brooks, C. S., and Scola, D. A. J. Colloid and Interface Sci. *32*, 561 (1970).
106. Belinski, C., Diot, C., and Keraly, F. X. L. Compt. Rend. C271, 1025 (1970).
107. Economy, J., and Lin, R. Y. Appl. Polymer Symposia no. 29, 199 (1976).
108. Aron, G. N., MacNair, R. N., and Erickson, R. L. Text. Res. J. *43*, 539 (1973).
109. MacNair, R. N., Aron, G. N., and Coffin, L. G. 12th Bienn. Conf. on CArbon, Pittsburgh, Penn., 1975, extended abstracts, p. 285.
110. Grindstaff, T. H. Text. Res. J. *39*, 958 (1969).
111. Princen, H. M. Text. Res. J. *40*, 1069 (1970).
112. Dyba, R. V., and Miller, B. Text. Res. J. *40*, 884 (1970).
113. Kemmel, J. M., and Steiger, F. M. 58th Amer. Chem. Soc. Natl. Meeting, New York, 1969.
114. Washburn, E. W. Phys. Rev. *17*, 273 (1921).
115. Chwastiak, S. J. Colloid and Interface Sci. *42*, 298 (1973).
116. Donnet, J. B., Cazeneuve, C., Schultz, J., and Shanahan, M. E. R. *14th Bienn. Conf. on Carbon*, Pennsylvania State University, University Park, Penn., 1979, extended abstracts, p. 216.
117. Cazeneuve, C. Docteur-Ingenieur Thesis, The University of Haute- Alsace and The University Louis Pasteur of Strasbourg, 1980.
118. Fitzer, E., Geigel, K. H., and Manocha, L. M. *5th London Intern. Carbon and Graphite Conf.*, vol. 1, Soc. of Chem. Ind., London, 1978, p. 405.
119. Bansal, R. C., and Chhabra, P. Unpublished data, 1981.

120. Yamamoto, M., Yamada, S., Sakatani, Y., Taguchi, M., and Yamagushi, Y. *Carbon Fibres—Their Composites and Applications*, The Plastics Institute, London, 1971, p. 179.
121. Herinckx, C., Perret, R., and Ruland, W. Nature *220*, 63 (1968).
122. Herinckx, C., Perret, R., and Ruland, W. *9th Bienn. Conf. on Carbon*, Boson, Mass., 1969, summary of papers, p. 121.
123. Herinckx, C., Perret, R., and Ruland, W. Carbon *10*, 711 (1972).
124. Hooley, J. G., and Dietz, V. R. Carbon *16*, 251 (1978).
125. Holley, J. G., and Dietz, V. R. *13th Bienn. Conf. on Carbon*, Irvine, Calif., 1977, extended abstracts, p. 301.
126. Raedersdorf, A. Ingénieur, Dissertation Diploma, C. N. A. M., Centre Régional Associé de Mulhouse, France.
127. Vogel, F. L. *Proc. 4th Intern. Carbon and Graphite Conf. 1974*, Soc. Chem. Ind., London, 1976, p. 332.
128. Vogel, F. L., and Popowich, R. *Petroleum Derived Carbons*, Am. Chem. Soc. Symposium Series No. 21 Am. Chem. Soc. 1975, p. 411.
129. Vogel, F. L. Carbon *14*, 175 (1976).
130. Fitzer, E., Geigel, K. H., Hüttner, W., and Weiss, R. Carbon *18*, 389 (1980).
131. Fitzer, E., and Weiss, R. Die Haftung von Kohlenstoffasern in Polymermatrix auf Grund chemischer Wechselwirkungen, Konstanz, April 1980, Vortragstagung DGM Verbundwerkstoffe, Deutsche Gesellschaft für Metallkunde (1981), S. 181–196.
132. Fitzer, E., Geigel, K. H., Heym, M., and Simitzis, J. Benetzung und Haftung zwischen Polymermatrix und Verstärk-ungsfasern Synopse und Mikrofiche, CIT 47 (1975), Heft 19, S. 811.
133. Donnet, J. B., Carbon, *20*, 267, (1982).
134. Donnet, J. B., and Ehrburger, P. *Proc. 4th Int. Carbon and Graphite Conf.*, Soc. Chem. Ind., London, 1976, p. 384; Carbon *15*, 143 (1977).
135. Dauksch, H. Ph.D. Thesis, Mulhouse and Strasbourg (1972).
136. Donnet, J. B., Ehrburger, P., and Ilvoas, A. M. Bull. Soc. Chim. France *6*, 2239 (1972).
137. Donnet, J. B. Carbon *20*, 268 (1982).
138. Fitzer, E. and Weiss, R. Intern. Symp. Univ. Liege, vol. 16–20, 1983. Society of Plastics Engineers.
139. Yamamoto, H. Intern. Symp. on Carbon, Toyohashi, Japan, 1982, p. 292.

4

Mechanical and Other Physical Properties

Aerospace structures need lightweight but strong materials that can withstand heavy loadings with minimum deflection. Thus the aerospace industry has generated a considerable amount of interest in developing materials with high specific strength and specific modulus. Several materials such as aluminum, titanium, beryllium, glass, boron, carbon, and so on have been examined, but most of these have been found to be brittle for such use. In order to minimize their brittleness, they have been obtained in the fibrous form for incorporation into metallic and polymer matrices. The specific strength and the moduli of some of these fibers and their epoxy resin composites are compared in Table 1. It is evident that beryllium fibers have low specific strength while glass fibers show a very low specific modulus. Boron and carbon fibers both exhibit a high specific strength and a high specific modulus. The strength and modulus properties of these fiber composites with epoxy resin matrix also indicate that they have more potential for use as high-performance materials in composites.

Carbon fibers heat-treated at high temperatures consist of graphitic crystallites with their axis preferentially oriented along the fiber axis. Since the electrical conductivity of graphite is considerably high in this crystallographic direction, and since this conductivity can be considerably enhanced by forming intercalation compounds, it should be possible to develop conducting carbon fibers suitable for electrical equipment. This has led to an upsurge in studies relating to the electronic and magnetic properties of carbon fibers and their intercalated compounds.

Carbon fibers are used in the form of composites which have an organic binder or matrix to weld them into a solid body. The carbon fiber contributes toward stiffness and strength and causes a large reduction in thermal expansion coefficient, a vast improvement in fatigue strength, and a decrease in electrical resistivity. The composites can be prepared by combining carbon fibers with other engineering materials in suitable proportions, by addition to different resin matrices, and by hybridizing with other fibers. With proper fabrication and placement of fibers and by controlling local concentration of fibers, the composites can be stronger and stiffer than

Table 1 Specific Strengths and Moduli of Some Fibers and Their Epoxy Composites

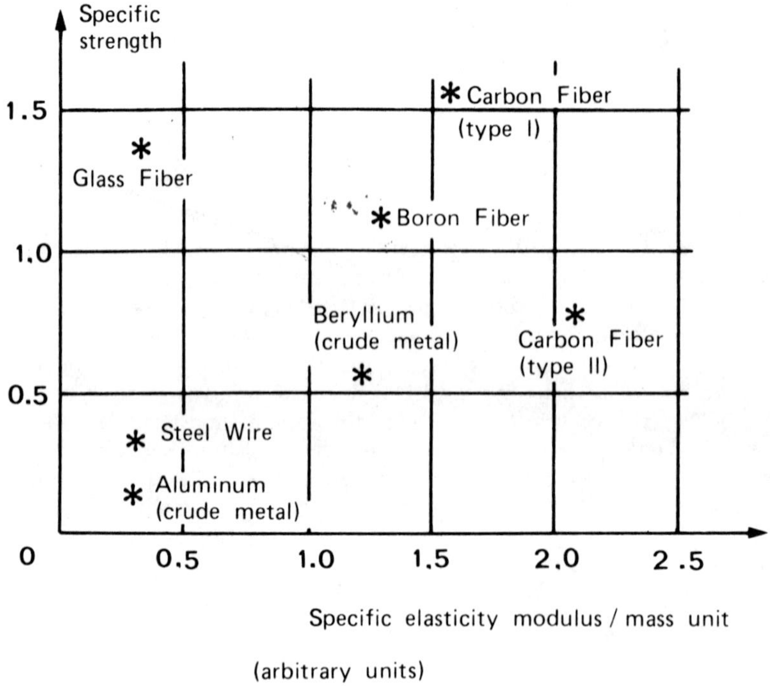

equivalent steel parts and weigh 40–70% less. Direct weight savings of these structures can, in turn, lead to additional indirect weight savings by requiring lighter supports. The composites when prepared with a suitable matrix system can have excellent environmental resistance and retain useful properties for long periods at temperatures above 300°C. Since carbon is very resistant to stress-corrosion deterioration, carbon-epoxy composites have replaced glass reinforced composites in alkaline environments.

Electrical conductivity of carbon fibers incorporated into a plastic matrix provides the benefits of static drain and RT (radio tracking) suppression in applications such as electronic equipment enclosures, automotive ignition devices, and small-appliance housing. The property is also useful in providing electrostatic paintability. Thermal conductivity enhances heat dissipation in components such as gears, brake pads, bearings, and other friction-related products, and in cryogenic processing equipment. Low value of linear thermal expansion coefficient of the carbon fiber in composites makes it possible to design structures with zero or very low linear or planar thermal expansion. This is useful in precision instruments such as telescopes and for the alignment of aerospace antenna. It may, however, be mentioned that the coefficient of thermal expansion of a composite depends not only on the orientation of the carbon fiber, but also on the thermal behavior of the matrix material. Thus, care must be taken to ensure that the volumetric expansion of the composite in the unrestrained direction does not exceed a tolerable limit. If such expansion is troublesome, changing to randomly oriented chopped fibers may be an acceptable solution.

The expansion of matrix material is greater than the expansion of carbon fibers. This can result in thermally induced stresses in composites. However, these stresses can be absorbed by the resin matrices since they are ductile and have larger elastic strain regions. Thus, the properties of the composites are not affected as long as the fiber—matrix interfacial bond strength is not exceeded.

The present chapter describes some of the mechanical, electronic, magnetic, and thermal properties of carbon fibers in relation to the type of precursor material and its processing conditions.

4.1 YOUNG'S MODULUS

The Young's modulus of a carbon fiber is one of the important mechanical properties which determine its use as a high-strength material. Moreover, the application of a carbon fiber in a composite depends very much on the translation of this property in the composite form.

4.1.1 Measurement of Young's Modulus

The Young's modulus is determined by using either a single fiber or a yarn of fiber filaments, usually with equipment which will enable a constant rate of strain to be applied.

In the case of single-filament determination, the filament is mounted on a heavy paper mounting tab containing a window of length equal to the desired gauge length, for example one inch [1]. The tab is then placed within the grips of the test machine and the two ends are separated by cutting or burning through both sides of the tab. The slope of elongation-load trace gives the apparent compliance of the filament. The compliance of the filament is calculated by subtracting the compliance of the machine from the apparent compliance.

The cross-sectional area of the filament is usually determined by measuring the diameter of the filament directly, using a microscope fitted with a split-image eye piece. However, this method is accurate only for filaments with circular cross-sections. Another method involves embedding the sample of parallel filaments in epoxy resin and obtaining a photomicrograph of the bundle cross-section. The area of cross-section of the individual filament is then determined using a polar planimeter. The cross-sectional area can also be obtained from the fiber density and the fiber weight per unit length.

It should, however, be mentioned that since the precise measurements of cross-sectional area involve several difficulties, it is usually taken as an average of several measurements.

When the Young's modulus is determined by using a yarn of fibers, the yarn is impregnated in epoxy resin, which is cured while the yarn is held straight on a frame. The impregnated yarn is then mounted on the testing machine with a proper alignment. A gauge length of about 10 in. is generally used in this method. The compliance is then determined from the elongation-load trace as in the case of the single-filament procedure. The cross-sectional area of the yarn is determined from the fiber density and the weight per unit length. As the resin modulus is only approximately 1% that of the fiber, the modulus of the resin is generally ignored [1].

Since the yarn method involves a larger amount of the sample, it is more representative and more relevant to the properties of the fiber when incorporated into a composite.

4.1.2 Young's Modulus and Preferred Orientation

Young's modulus is an intrinsic property and is governed by the orientation of the graphitic crystallites relative to the fiber axis. Thus, a sample of carbon fiber which had basal planes of the crystallites lying within ± 35° to the fiber axis showed a Young's modulus of 103 GN m^2, while another sample with an orientation angle of ± 10° was found to have a Young's modulus of 410 GN m^2. Therefore, any process which could reduce this angle of crystallite orientation relative to the fiber axis (i.e., which enhances the preferred orientation of the graphitic crystallites), results in enhancing the Young's modulus.

Fourdeux and co-workers [2] measured the orientation of a number of rayon-, PAN-, and pitch-based carbon fibers in terms of the parameter q, and observed that, except for some partially graphitized carbon fibers whose moduli were lower than the theoretical values, the modulii of carbon fibers seemed to be a function of the degree of preferred orientation. The relationship between the preferred orientation and Young's modulus was explained on the basis of the wrinkled-ribbon model with two orientation-independent compliance parameters corresponding to the compliance of the hexagonal carbon network within its plane (S_{11}), and the compliance corresponding to the environmental resistance against tilting of the layer planes (k). The Young's modulus (E_c) is thus given by

$$\frac{1}{E_c} = \ell_z S_{11} + m_z k \qquad (1)$$

where

$$\ell_z = \frac{\int \sin^2\phi \; I(\phi) \; d\phi}{\int \sin\phi \; I(\phi) \; d\phi} \tag{2}$$

and

$$m_z = \frac{\int \cos^2\phi \; I(\phi) \; d\phi}{\int \sin\phi \; I(\phi) \; d\phi} \tag{3}$$

The parameters ℓ_z and m_z are calculated from the orientation distribution function $I(\phi)$, where ϕ is the angle between the layer normal and the fiber axis. Both ℓ_z and m_z can be expressed as functions of q as well.

The results obtained by the wrinkled-ribbon model were compared with those from uniform stress or uniform strain models [3], and were observed to fit the experimental values better for a wide range of fiber modulii, heat treatment temperatures, and precursors. The uniform stress and uniform strain models required the use of shear compliances which varied with the preferred orientation. Small deviations between the experimental and theoretical values in the case of high-modulus rayon fibers were attributed to the lower shear modulus in more perfect carbon fibers [3].

Tokarsky and Diefendorf [4] also observed significant deviations from the Fourdeux plots in the case of many PAN carbon fibers. In their case, they found that the experimental values were higher than those predicted from the preferred orientation parameter q. They believe that the deviations are not due to variation in the shear modulus, but to the gradient in preferred orientation through the cross-section of the fiber. According to them, carbon fibers have a composite structure (dog-bone shaped) with almost perfectly axially oriented material existing close to the surface, while poorly oriented material is present in the core and the lobes of the dog bone. These gradients give rise to a constant strain component, so that if moduli are to be calculated from the elastic constants, not only preferred orientation but these gradients in preferred orientation must also be known. Stevens and co-workers [5] recently determined Young's modulus of carbon fibers after abrasion of the surface by oxidation or ion milling and actually observed a modulus gradient within the fiber, with the surface layers having a higher modulus and the core material a lower modulus. The origin of this modulus gradient has been related to relaxation during the initial stabilization stage of carbon fiber processing of the oriented PAN precursor fiber. Highly aligned molecules produced during stretching of the precursor fiber remain aligned only at the surface owing to cyclization. However, molecules near the center of the fiber stabilize more slowly due to a decreased oxygen concentration [6], and are more likely to relax, forming a more poorly aligned structure at the interior of the carbon fiber. The alignment at the surface is retained through the carbonization process to produce high modulus at the surface of the carbon fiber, while the poorly aligned material results in lower modulus in the core.

Jones and Duncan [7] found that the mean modulus of type I PAN carbon fibers decreased from 360 GN m^2 at a diameter of 7.3 μm to about 300 GN m^2 at a diameter of 9.4 μm while Perry et al. [8] and Delamotte and Perry [9] observed a similar fall from 290 GN m^2 at 7.5 μm to 250 GN m^2 at 10.5 μm diameter for type II carbon fibers. This decrease in Young's modulus with an increase in diameter has been attributed [7] to the existence of a dual structure, as suggested by Diefendorf and Tokarsky [4] and Stevens et al. [5]. The greater portion of thin fibers will be composed of a sheath derived from the oxidized portion of the polymer, while a thick fiber will contain a large fraction of the core. Since the outer sheath has a better alignment of the crystallites, thin fibers would exhibit a higher modulus than the thick fibers.

4.1.3 Influence of Stretch on Young's Modulus

The preferred orientation of a carbon fiber can be enhanced by stretching the fiber during one of the processing stages. This stretch orientation is usually carried out during the low-temperature heat treatment in the case of PAN carbon fibers; during graphitization in the case of rayon; and during spinning of the pitch fibers. Stretch graphitization improves the Young's modulus of carbon fibers when carried out at high temperature, but is an expensive and difficult process and, therefore, seems to be frequently avoided in the case of PAN and pitch carbon fibers.

In rayon carbon fibers, however, stretch graphitization is essential, since stretching during low-temperature heat treatment or during the carbonization stages does not appreciably affect the preferred orientation [1]. This is due to the fact that during the low-temperature heat treatment (≅ 300°C) the cellulosic structure is undergoing decomposition and degradation reaction. Although stretching during carbonization improves the degree of preferred orientation, the effect is much more pronounced when stretching is carried out during graphitization. For example, stretch carbonization could raise the Young's modulus from 42 to 70 GN m^2, whereas the stretch graphitization for the same sample of rayon fibers raised the Young's modulus to 210 GN m^2 [1], although the stretch during graphitization was 50% of the stretch during carbonization. Bacon and Schalamon [10], and Gibson and Langlois [11] observed that the modulus of a rayon carbon fiber could be controlled by the amount of stretch. Rayon carbon fibers with a modulus of 620 GN m^2 were prepared by stretching up to 400% [10]. However, PAN carbon fibers with a similar value of Young's modulus need only a 29% stretch at 2800°C. The high cost of stretch graphitization results in limited and specialized use of rayon carbon fibers.

The effect of stretching during the preoxidation of PAN fibers has been extensively studied [12–15]. It was observed that the tows of carbon fibers which were allowed to shrink during the oxidation treatment produced carbon fibers with lower Young's modulus while those which were elongated gave carbon fibers with high Young's moduli. The Young's modulus increased with an increase in the degree of elongation. The stretching is believed to render the molecular chains more parallel to the fiber axis during the cross-linking reaction, making the graphitic crystallites more oriented. Moreton [16] observed that steam stretching of PAN fibers, before and after carbonization, also enhanced the Young's modulus. The degree of enhancement depended upon the steam–stretch ratio.

High-modulus carbon fibers have also been obtained from pitch precursor fibers by hot stretching [17,18]. Baker et al. [17] obtained carbon fibers with a Young's modulus of 440 GN m^2 by an elongation of 120% at temperatures higher than 2000°C.

However, it has been reported recently by Singer [129] that mesophase pitch-based carbon fiber having elastic moduli ranging from 175 to 700 GN m^2 can now be produced (the theoretical maximum being about 1000 GN m^2) without stretching after heat treatment at high temperatures. The fiber structure is more graphitizable than in the case of rayon- or PAN-based fibers and the final moduli and high-tensile strength obtained depend on the concentration of various types of structural defects. Stretching is of importance to enhance graphitization and the preferred orientation.

4.1.4 Influence of HTT on Young's Modulus

Preferred orientation in carbon fibers can also be enhanced by heat treatment at high temperatures. Consequently, the effect of heat treatment temperature (HTT) on the Young's modulus has been investigated by a number of workers [1,2,19,20]. It has been observed that the Young's modulus invariably increases with an increase in the heat treatment temperature (Table 2). Bright and Singer [20] have recently studied the Young's modulus of mesophase pitch carbon fibers with random and radial structures and observed that the Young's modulus increased with HTT, but the radial structured fibers generally gave higher modulus values for the same HTT. This has been attributed to better alignment of the crystallites in radial fibers compared to the random structured fibers.

Table 2 Young's Modulus of PAN Carbon Fibers Heat Treated at Different Temperatures

Temperature (°C)	Modulus ($GN\ m^{-2}$)
1000	214
1250	242
1400	248
1600	248
1800	283
2000	338
2200	352
2400	386

Data after Ref. 38.

4.1.5 Influence of Boron Doping and Neutron Irradiation on Young's Modulus

Addition of boron to carbon fibers increases the Young's modulus [21]. The increase depends upon the temperature of annealing and the amount of boron. In the case of type I PAN carbon fibers the Young's modulus increased from 415 to 500 GN m^2 with 1% boron at an annealing temperature of 2750°C [21]. The presence of boron tends to increase the rate or degree of recrystallization of the carbon fibers at any particular annealing temperature, which enhances the degree of preferred orientation and perfection of the crystallites.

Irradiation with neutron increased the Young's modulus of both type I and type II carbon fibers, but only by about 10% (Table 3). This was attributed to an increase in shear modulus of the graphite parallel to the basal planes [21]. Wicks [22] investigated the effect of fast neutron irradiation on the Young's modulus of carbon fibers heat-treated to different temperatures, and observed that the Young's modulus increased initially, but a prolonged exposure resulted in degradation of the fiber, causing a lowering of the Young's modulus. At the highest dose the modulus decreased to a value lower than that for the unirradiated fiber. The dose at which the maximum value of modulus was obtained increased with increasing HTT. The increase in modulus at low doses has been attributed to a decrease in the mobility of basal dislocations by point defect damage, which reduces the plastic flow at the dip of microstructural stress concentrators, thus decreasing the effectiveness of a fracture. At higher doses the Young's modulus decreases due to the increase in dislocation density and misalignment of graphite layers [22].

Table 3 Effect of Irradiation on Young's Modulus of Carbon Fibers

Carbon fiber	Treatment	Young's modulus (GN m^2)
Type I	Unirradiated	358
	Irradiated	396
Type II	Unirradiated	256
	Irradiated	284

Source: Data from Ref. 21.

4.2 TORSIONAL MODULUS

The torsional modulus of a carbon fiber is usually determined by a torsion pendulum. The one used by Henricksen and Fischbach [23] consisted of an inertial bob composed of a pair of identical small lightweight mirrors enclosing a short segment of the fiber, which was also the armature of an electrostatic excitation system. The bob was attached at the midpoint of a single fiber (about 2.5 cm long), mounted vertically between two supports with an epoxy cement. The upper support was attached to one end of the beam of a modified chainomatic balance permitting tensile loading over a range of several grams. The lower support was attached to a shaft coaxial with the fiber so that static torsional strain could be applied. The pendulum was operated in vacuum to avoid air damping and was mounted on a heavy vibration isolating base. The torsional modulus (G) was determined from the moment of inertia of the bob (I), the fiber dimensions, and the period of oscillation (Γ) using the equation

$$G = \frac{2\pi I \ell}{A^2 \Gamma^2} \tag{4}$$

where ℓ is the length of the fiber between the bob and the end supports (upper plus lower portions), and A is its cross-sectional area.

Hawthorne [24] observed that the shear modulus in the case of petroleum pitch carbon fibers decreased with increased stress elongation. The value decreased from ≅ 16 GPa to ≅ 10 GPa with a stress elongation variation from 0 to 150%. The shear modulus was lower than that of PAN carbon fibers [25], which is probably due to differences in the microstructures of the two fibers. Adams [26], however, observed that the shear modulus was almost independent of hot

Table 4 Torsional and Young's Modulus of Different Carbon Fiber Types

Carbon fiber type	Young's modulus (GN m^2)	Torsional modulus (GN m^2)
Pitch mesophase		
PM-A	184	13.5
PM-B	262	9.0
		9.2
PM-C	364	10.8
		9.7
PM-C (2700)	400	10.4
Isotropic pitch		
KCF-200	80	16.4
KCF-2700	∿80	18.0
PAN		
T-400	226	21.4
A-S	215	21.0
HM-S(H)	∿ 370	20.2
Modmor I	∿ 400	28.2
HM-S(C)	∿ 380	35.3
Rayon		
T-11 (carbon)	79	15.3
T-12 (graphite)	83	22.8
T-25	176	12.3
T-70	390	12.3
T-75	540	13.8
T-100	680	16.9

Source: Data from Ref. 28.

stretching up to 180% and showed a value of ≅ 18 GPa. In the case of PAN carbon fibers the shear modulus was more or less independent of the HTT in the range 1000–2500°C but showed a slight decrease at higher HTTs.

Fischbach [27] and others [23,28,29] studied the dynamic torsional behavior of a number of carbon fibers from PAN, rayon, isotropic pitch, and mesophase pitch precursors with varying Young's modulus, and observed that the value of torsional modulus was almost one order of magnitude lower than the Young's modulus. This low torsional modulus is inconsistent with the intrinsic anisotropy of the carbon fiber material. In the case of rayon and pitch carbon fibers, the torsional modulus increased only by about 50% (from 12 to 18 GPa) when the HTT was increased from 1000 to 2700°C (Table 4), and was relatively insensitive to Young's modulus and the longitudinal orientation texture. The torsional modulus for PAN carbon fibers was, however, larger than those of rayon or pitch carbon fibers and showed a variation between 18 and 35 GPa, and the value was found to be more sensitive to the precursor or processing conditions. The torsional modulus for mesophase pitch carbon fibers was even lower than for rayon carbon fibers. These variations in the torsional modulus in different types of carbon fibers may be attributed partly to the differences in microstructural morphology and cross-sectional shape. The most important factor, according to Fischbach [27], is the transverse layer orientation texture in the cross-section. Mesophase pitch carbon fibers are known to possess an appreciable radial structure in which interlayer shear dominates the torsional stiffness, and consequently, these carbon fibers display the lowest value of torsional modulus. PAN carbon fibers contain strong circumferential textures [30] where interlayer stiffness is a contributing factor and, therefore, show high torsional modulus.

4.3 STRENGTH OF CARBON FIBERS

The strength of carbon fibers is a complex field of study and is investigated under tension, bending, compression, and torsional shear. The tensile strength is measured by carrying the tensile test for modulus to failure. The strength under bending is usually measured by the Sinclair loop test [31], in which a single filament, shaped into a loop, is confined to a vertical plane between two glass slides lubricated with glycerine. The filament is supported at the upper end and is stressed by weights fastened at the lower end. The loop diameter is measured as a function of weight using a telescope. The strength σ is then calculated by using the equation [32]

$$\sigma = 16(w) \frac{L}{\pi d^3} \tag{5}$$

where w is the applied weight and L = 3.75 D, where D is diameter of the loop, and d is filament diameter.

The compressive strength is measured [33] by embedding the filament in a clear-resin tensile specimen. The filament is oriented in a direction transverse to the resin specimen's tensile axis. The application of tensile load subjects the filament to compressive forces that ultimately result in its failure.

Torsional shear strength is usually measured by carrying to failure the static torsion test for shear modulus [34].

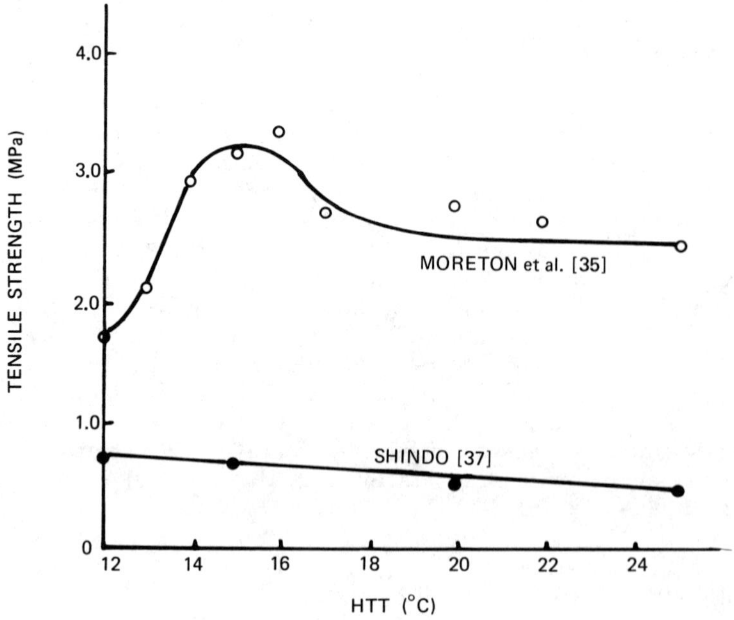

Figure 1 Strength of carbon fibers as a function of heat treatment temperature. (From Ref. 35.)

It is apparent from the methods of measurements of strength that the majority of the strength data are generally obtained from tests carried out on single filaments. Consequently the results show a wide variation. In many cases this variation has been attributed to the existence of flaws in the carbon fiber. But this scattering of data, which has been found to be as much as 25%, can also result (at least partly) from several shortcomings in measurement techniques. One of these limitations could be the difficulty in mounting single filaments on the tensile machine without damage to the filament, and in such a way that only tensile load is applied to it. Another limitation is the precise measurement of the fiber diameter. The carbon fiber may not have a uniform diameter throughout its length and all the fibers in a tow may not have exactly the same diameter. Therefore, it is often difficult to compare the strengths of a particular brand of carbon fiber, determined in different laboratories.

The strength, like the modulus of a carbon fiber is not an intrinsic property and depends upon the type of the precursor fiber, processing conditions, such as heat treatment temperature (HTT), preoxidation conditions, and the presence of flaws and defects.

Table 5 Strength of Mesophase Pitch Carbon Fibers as a Function of HTT

Carbon fiber	HTT (°C)	Strength (GN m^2)
Radial	1000	0.956
2-stage	1700	1.48
Thermosetting	2000	1.38
	2500	1.32
	3000	2.30
Random	1000	0.996
2-stage	1700	2.01
Thermosetting	2000	1.51
	2500	2.01
	3000	2.03

Source: Data from Ref. 20.

4.3.1 Effect of HTT on Strength

It has been observed [12,35,36] that the strength of PAN carbon fibers increases upon heat treatment, reaches a maximum value at about 1500°C, and decreases thereafter at higher heat treatment temperatures. Shindo [37], on the other hand, observed no such increase in strength on HTT (Figure 1). He rather observed a slight decrease which he attributed to a change in the microstructure of the carbon fiber. This, however, seems very unlikely since increased HTT increases the orientation of the carbon fiber. Johnson and Tyson [38] found that the strength was maximum at 1250°C and decreased thereafter. The decrease in strength was considerably larger beyond a HTT of 1800°C.

Bright and Singer [20] examined the effect of HTT on the strength of mesophase pitch carbon fibers with random and radial structures and thermoset in a single- and a two-stage process. They observed (Table 5) that the tensile strengths of random structure fibers were higher at low HTT than those of the radial structure fibers. However, the random fibers did not improve in strength above 2500°C HTT while the radial fibers improved in strength sharply and became stronger than the random fibers. Fibers thermoset in a two-stage process were stronger than the carbon fiber thermoset in a single-stage process with the same structure. In all cases the strength decreased as the HTT increased from 1700 to 2000°C, but increased again at higher HTTs. The decrease in strength has been attributed to the probable contamination of the carbon fiber with silicon or other impurities diffusing into the fiber from the furnace. The difference in behavior of the random and radial structure fibers above 2500°C HTT is suggested to be due to differences in the ease of graphitizability of the two structures.

The decrease in strength of carbon fibers above 1700°C has also been attributed [38] to the elimination, at high temperatures, of amorphous carbon which is responsible for cross-linking between the small crystallites involving crystallite edge atoms and interstitial atoms. As the HTT is raised the cross-linking decreases, making the layer planes more mobile and resulting in a more perfect order of the graphitic planes. Thus, below 1900°C, strength is a function of the degree of cross-linking. Above 1900°C the strength remains more or less unchanged [38] since the degree of cross-linking is almost negligible.

Jones and Duncan [7], however, feel that the breaking of cross-links between crystallites cannot result in a simultaneous increase in modulus and a decrease in strength when the fibers are heated above 1500°C. According to them, high-temperature heat treatment can cause a specific type of flaw in the carbon fiber, which may be visualized as a basal-plane crack formed during cooling in a manner similar to the formation of Mrozowski cracks in bulk graphite [39]. Sharp and Burnay [40] are of the view that the decrease in strength beyond 1800°C is due to the presence of inclusions in the carbon fiber which react and volatilize at these higher temperatures to form voids. These voids result in the formation of misoriented three-dimensional graphite on the walls of the fiber.

Le Maistre [41], working with dog bone cross-section fibers and Moreton [42], working with circular fibers spun in a clean atmosphere, have both found that tensile strength can continue to rise up to 2800°C HTT. The flaws undoubtedly play a part in determining the strength achieved in single fibers, particularly at HTT up to 1500°C [43–45], as can be shown by etching away the surface. This decrease in number of surface flaws, while not affecting the internal flaws, raises the potential strength at low HTTs but does not affect the strength at higher HTTs. Le Maistre [41] has suggested that the flaws responsible for the decrease in strength at high HTT are cracks induced by residual stresses resulting from cooling of a fiber with roughly concentric graphitic layers. This effect should be less in a dogbone-shaped carbon fiber with two lobes of effectively smaller diameter. Le Maistre observed that for his PAN-based fibers with dogbone cross-section the tensile strength, in fact, continued to rise up to at least 2500°C.

4.3.2 Effect of Preoxidation Conditions on Strength

Watt and Johnson [12] studied the effect of length changes of PAN fibers during preoxidation at 220°C on the strength of carbon fibers and found that the strengths were lower in those fibers which were allowed to shrink during oxidation than for those which had been stretched. Since tensile strength is not an intrinsic property, but is dependent upon the presence, nature, and number of structural defects, it appears the shrinking of the fibers introduces defects. These could arise from an increase in any amorphous material present in the PAN precursor material, which produces much less-ordered carbon and can be a source of weakness. However, increasing the load beyond a certain optimum value decreased the strength of the resulting carbon fibers [46]. This has been attributed to sudden elongation of the fiber, which resulted in a bond rupture, and consequently, in a decrease in strength. Moreton [16] prepared PAN fibers by wet spinning under varying degress of stretch (5—13 times) and converted them into carbon fibers by normal procedures of heat treatment up to 2500°C. The strength of the carbon fibers increased with the degree of steam stretch and was maximum with a stretch of 13 times. Fitzer and Müller [47,48] observed that the strength of the carbon fibers increased considerably and continuously with the degree of prestretching of the textile precursor fiber. This has been attributed to the variation of the rate of cyclization reaction. They studied the kinetics of the cyclization rate with regard to the degree of prestretch using DTA and observed that the cyclization rate decreased with increase in textile prestretching [48]. In addition, the degree of prestretching was related to decreases in cross-sectional area of the fiber, giving a higher specific surface which resulted in a higher absorbance of oxygen, which inhibits the cyclization rate. It was previously known that cyclization depended upon oxygen uptake. This study showed for the first time that cyclization depends upon the degree of prestretching of the fiber.

Strength was also found to be influenced by the nature of the pretreatment atmosphere, the temperature, and the length of time of the pretreatment [12,13,46,49]. Heating in air at 220°C up to 22 hr or more caused significant increases in strength, although heat treatment in nitrogen at the same temperature lowered the strength [12]. The treatment in air for 2 hr produced higher strength carbon fibers than treatment with nitrogen for 16 hr, even though air-treated fibers were found to exhibit a bubble effect. Similarly, treatment in air at 215°C produced stronger fibers than treatment in oxygen. This is due in part to the damaging effect of oxygen [46]. Strength increased with an increase in the degree of aromatization, which results from preoxidation. Preoxidation stabilizes the molecular chains and prevents them from cleavage during carbonization. However, if oxidation is carried out for too long a period, the strength decreases [50].

4.3.3 Effect of Fiber Diameter on Strength

The tensile strength of carbon fibers has been found to vary inversely with the diameter of the fiber [7,8,9,51]. Since the tensile strength is influenced by the presence of flaws, it could be immediately inferred that the variation in strength with fiber diameter should be explained as a volume effect, since more flaws are expected to be present in a unit length of thick fiber than in a thin fiber. However, Jones and Duncan [7] observed no relationship between strength and fiber diameter for PAN-based carbon fibers heat-treated to about 1200°C, although a relationship was found when the carbon fibers were heat-treated to higher temperatures. Since heat treatment at higher temperatures is known to cause Mrozowski cracks, it is quite probable that the decrease in strength is due to them. As these basal-plane cracks are formed from anisotropic thermal contraction of the crystallites, a considerably larger number of them would occur in randomly oriented crystallites. It is now well established that carbon fibers have a composite structure with a sheath consisting of well-oriented crystallites and a core comprised of less-oriented material. In the case of thick fibers the core constitutes a larger fraction of the volume than in the case of thin fibers. Consequently, more cracks are likely to occur in a thick fiber. In addition, misorientation of layer planes is larger in the core than in the sheath, so that a larger number of cracks in thicker fibers will be oriented at a significant angle to the fiber axis. Since such cracks are likely to be more effective as flaws than cracks of equivalent size aligned close to the fiber axis, the thicker fibers would exhibit lower tensile strengths.

Perry et al. [8] and Delamotte and Perry [9] also observed a strong dependence of the mechanical properties of PAN carbon fibers on their diameter. This has been attributed partly to structural variations through the fibers which cause the outer regions to be stiffer, and partly to the presence of cracks or other local flaws. Both factors are likely to be more serious in thick fibers than in thin fibers.

Hsiung and Rowe [52] measured diameters and strength of 15 samples of three given lots of pitch-based carbon fibers and found that the fiber strength was almost independent of the fiber diaameter, although the fibers with the most uniform diameters possessed the highest strength. This is probably due to more even distribution of stresses in samples with more uniform size distributions.

4.3.4 Effect of Doping and Irradiation on Strength

Heat treatment at 2750°C in the presence of boron did not change the strength of type II carbon fibers, while the strength of high-modulus type I carbon fibers increased slightly when the boron content was about 1% [21]. Neutron irradiation, however, increased the tensile strength by about 10% [21,53,54]. The increase in strength on irradiation is due to the fact that it interferes with the dislocation motion without significantly altering the crystal structure. Upon heat treatment with boron, two opposing processes appear to take place. The presence of boron atoms increases crystallite diameter in both the c and a directions, which tends to decrease the strength and hinder the dislocation motion in the graphite lattice, thereby increasing the resistance to shear of the doped material. Consequently, the strength remains more or less unchanged. Jones and Peggs [55] observed a decrease in strength of PAN carbon fibers upon neutron irradiation. However, the irradiation was carried out at a high temperature of 420–500°C for a prolonged period of time, which makes interpretation of the results difficult. It has been suggested that the effect may be due more to oxidation than to radiation damage [56].

Wicks [22] investigated the influence of fast neutron irradiation on the strength of carbon fibers heat-treated to various stages of graphitization (1600, 2500, and 3000°C). The strength variations were significant for 1600 and 2500°C heat-treated fibers. All carbon fibers initially increased in strength at low and intermediate doses, but decreased in strength at high doses. The dose at which the maximum strength was obtained increased with the degree of graphitization. The variation in strength on irradiation has been attributed to changes in the energy required to propagate cracks from preexisting platelets of graphite [22].

Jackson and Marjoram [57,58] observed that the heat treatment of type II carbon fibers with nickel and cobalt at temperatures in the range 700–1100°C for prolonged periods of time, resulted in lowering the strength. The type I carbon fibers, however, maintained their strength up to 1000°C, but lost it thereafter. These workers calculated the activation energies of these processes and related them to the carbon diffusion into these metals. They are of the view that the stability of type I carbon fibers is due to their more highly crystallized surfaces with fewer exposed edge atoms. At higher temperature there is reprecipitation of the crystallites in good size, but their orientation and the continuity of the basal planes is lost, resulting in weakening of the fiber. These effects are responsible for the difficulties in preparing carbon fiber-nickel composites [59]. Barclay and Borfield [60] found that pure nickel vapors react with carbon fibers at 1080°C and decrease their strength, but without recrystallization.

4.4 DEFECTS AND FLAWS IN CARBON FIBERS

Almost all solids are associated with a certain degree of structural imperfection and flaws which create internal stress in the material. In brittle materials, fracture is generally operative at these structural faults or flaws and, therefore, the strength is usually well below the intrinsic strength given by the equation

$$\sigma = \frac{\sigma_1}{k} \tag{6}$$

where σ_1 is the intrinsic strength, σ is the observed strength, and k is the stress concentration factor which is usually on the order of 10^2 or more.

Among the carbon materials, graphite whiskers are the strongest because they are almost completely free of flaws. Graphite whiskers with strengths as high as 21 GPa have been obtained. Carbon fibers, on the other hand, are always associated with a number of flaws and defects, including dislocations, stacking faults, metal particle inclusions, gross discontinuities, or damages of the pyrolyzing fiber. Consequently, lower strengths have been realized.

Johnson [61] was the first to indicate that tensile fracture in carbon fibers occurred at the site of flaws which may be present inside the fiber (internal flaws) or on the surface (surface flaws). Internal flaws are generally derived from the precursor polymer fiber and retain their geometrical identity through the carbon fiber processing cycle. Increase in the average strength of a carbon fiber with decreasing gauge length indicated that these strength-lowering defects or flaws are randomly distributed. Optical microscopic examination of the fibers showed [62] that some of the flaws are voids or inclusions in the parent fiber, which had apparently remained unchanged during the carbonization and graphitization processes, although material in them might have been lost by diffusion or reaction.

The strength distribution of flaws in a precursor polymer fiber was studied by optical microscopy and scanning electron microscopy (SEM) examination of fracture surfaces after etching the surfaces by ion bombardment in an ionized plasma of argon at reduced pressure or by air etching at 450°C [63,64]. It was found that the strength of each carbon fiber increased considerably on etching the surface and the fiber generally failed at the internal flaws. This indicates that internal flaws operate at higher strength distributions than the surface flaws. Consequently, these workers suggested three types of flaws with different strength distributions (Figure 2). Surface flaws generally predominate in as-prepared carbon fibers and are associated with fairly normal strength distributions (Figure 2a). Internal flaws are associated with somewhat higher strength distributions (Figure 2b) and operate either in the overlapping region of strength distribution or when the surface flaws are eliminated or modified to give higher strength distributions (Figure 2c). Therefore, upon elimination of the surface flaws by any etching technique, the mean strength of the

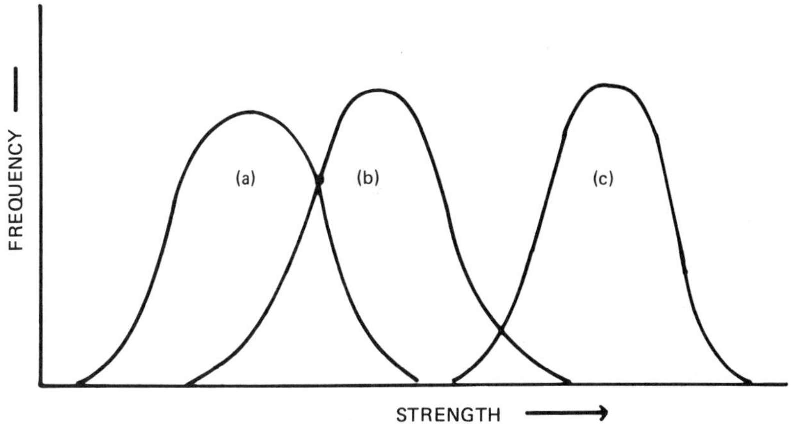

Figure 2 Strength distributions of different types of flaws. (a) Original surface flaws, (b) internal flaws, (c) surface flaws after etching. (From Ref. 64 with permission from Pergamon Press Inc., Great Britain.)

fiber is improved and the carbon fiber fails at the internal flaws. The importance of etching is, thus, the replacement of distribution "a" by distribution "c", allowing the effective operation of internal flaws by distribution "b."

The importance of flaws in influencing the strength of carbon fibers was also stressed by Moreton [65] in his experiments of strength measurements at different gauge lengths. He measured strengths of fibers in lengths of 0.5, 5.0, and 10.0 cm over a range of heat treatment temperatures and observed that the average strength decreased exponentially with length. This has been attributed to the presence of randomly occurring defects, which are likely to occur in greater number as the test length of the carbon fiber is increased. The strength of the fiber was found to be maximum at all gauge lengths at 1500°C. Chwastiak et al. [66] determined the tensile strengths of mesophase pitch carbon fibers as a function of gauge length in the range 1.4–40 mm. The filaments were examined from a fiber apparently free from flaws, and the results were compared with data from filaments weakened by internal voids. Filaments with 8 μm diameter were found to give strengths of 3.4 GPa at a gauge length of 1.4 mm. The strength data were analyzed using a two-parameter Weibull equation. The shapes of the probability of failure versus strength plots and the appearance of the histograms showed that a secondary flaw population became significant at longer filament lengths. Extrapolations of flaw distribution at short test gauge lengths gave a predicted

(a)

(b)

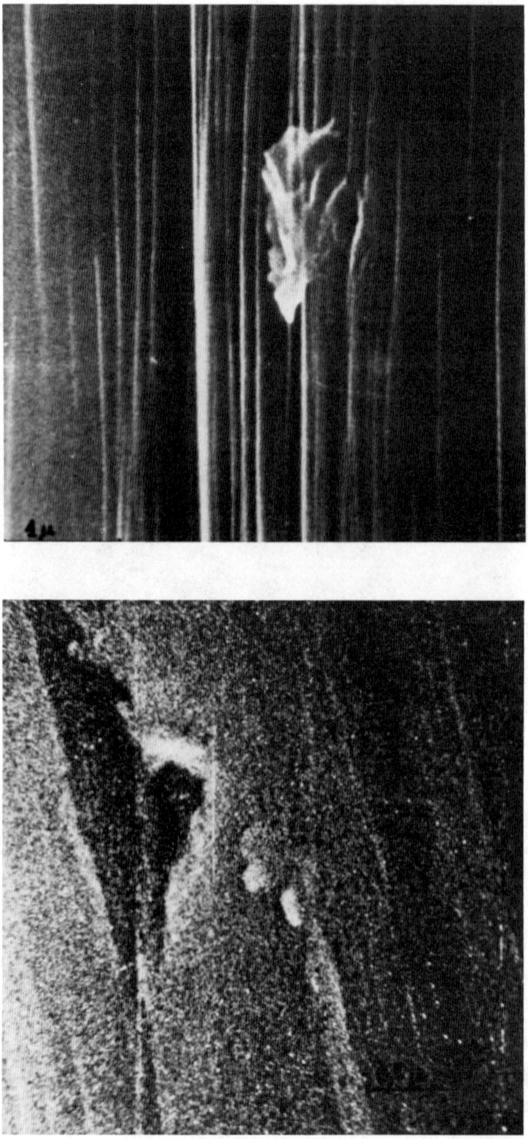

Figure 3 Sulfur containing surface flaws in type I and type II carbon fibers. (a) Surface flaw—fiber type I. (b) Surface flaw—fiber type II. (c) Surface defects as revealed after electrolyte oxidation. (d) The same as seen by Eddax revealing the sulfur-rich zones. (1 and 2 from Ref. 127. 3 from Ref. 126.)

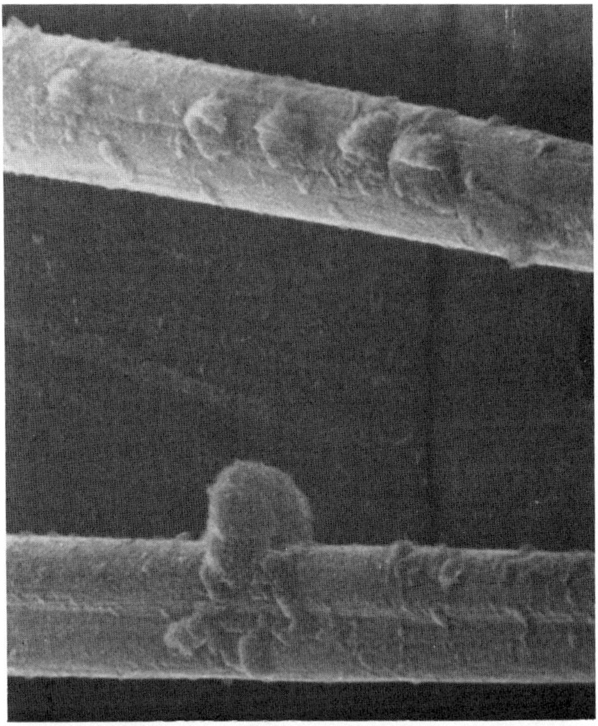

(c)

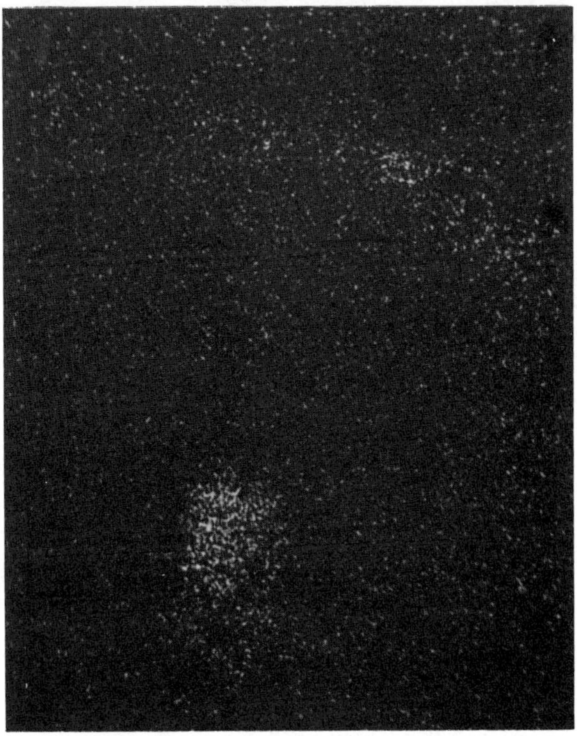

(d)

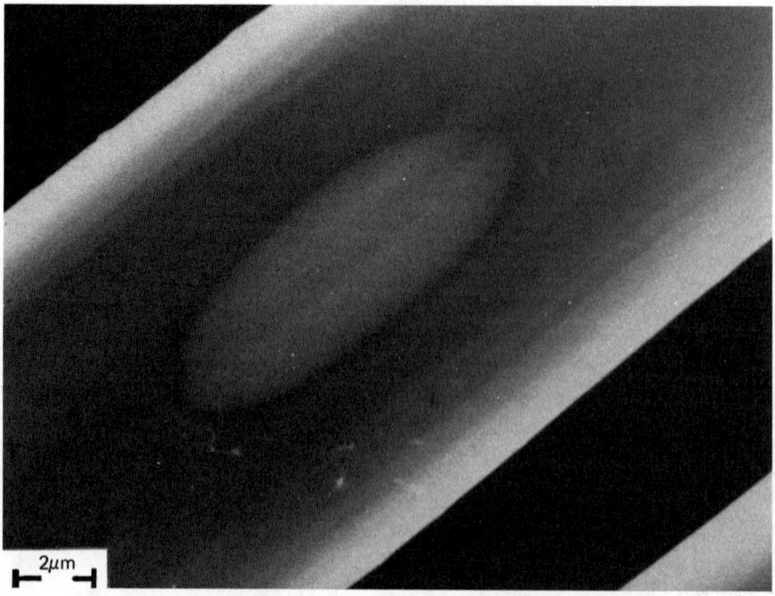

Figure 3e Central cavity. (From Ref. 68; photo courtesy of J. V. Sharp.)

strength of 7 GPa at 0.1 mm length for both flaw-free and flawed filaments, which is close to the maximum value reported for PAN fibers [67].

The majority of the internal flaws observed in prestretched acrylic fibers were recognized [63,64] as elongated voids or cavities of 1–4 μm in width. Prestretching of the fiber did not increase the concentration of the flaws, but simply highlighted their presence. The flaws were distributed extremely unevenly along the fiber length and approximated to a Poisson-type distribution, depending only on the flaw concentration per unit length. The flaw concentration for as-received fibers was in the range of 0.03–2.3 flaws per mm. However, this concentration could be reduced by suitable stretching of the fibers.

4.4.1 Types of Flaws

Broadly speaking, the flaws in carbon fibers could be classified into two types: (1) surface flaws and (2) internal flaws.

Surface Flaws

These include surface imperfections, fibril boundaries, and bends and kinks in the microfibrils or surface pits. These flaws control the strength of carbon fibers which have not been heat-treated beyond 1000–1200°C. Such fibers behave in a most brittle manner compared to carbon fibers heat-treated to higher temperatures. Bends and kinks in the microfibril tend to straighten out under longitudinal tension and tend to become more severe under longitudinal compression. These flaws are observed in most fibers, but little is known about their structure.

Typical surface defects from type I and type II PAN fibers are shown in Figure 3a-d and it has been shown recently by Messiet et al. [126] that electrolytic oxidation can reveal surface defects which are sulfur-containing probably from residues left by the precursor [127a].

Internal Flaws

Sharp and co-workers [68,69] examined a wide range of carbon fibers with circular and dogbone cross-sections heat-treated between 1000 and 2800°C by high-voltage electron microscopy after etching the fiber surface, and observed several different types of internal flaws which are described below.

Central cavities These cavities (Figure 3e) occur along the central axis of the fibers and are elongated along the fiber axis up to about 50 μm. The maximum diameter of such cavities is about 5 μm. Inclusions are occasionally observed in cavities, and in some cases there is a local increase in the fiber diameter at the cavity, which indicates that the cavities are probably formed by the expansion of gas at high temperatures, producing local plastic deformation of the fiber. These cavities reduce the strength of the fiber by reducing the effective cross-sectional area.

Diconical voids (Figure 4). These are present in almost all carbon fibers and are randomly distributed across the cross-section of the fiber. The shape and diameter vary with the type of the carbon fiber, but they are usually 1–3 μm in diameter. In some cases they are associated with small particulate inclusions. These voids are thought to be caused either by gases dissolved in the fiber-spinning solution or by irregular flow at the spinnerette orifice by the rapid coagulation of the polymer.

Inclusions (Figure 5). These generally have irregular circular shapes with a maximum size of about 1 μm in type II carbon fibers. In the case of certain pitch- and coal-based carbon fibers, larger inclusions have also been noted [70]. In some cases the inclusions may be associated with thin cavities on either side.

Needle-shaped cavities (Figure 6). These are extended narrow cavities with very small diameters of 0.1 μm. However, their lengths are usually around 5 μm. They are generally not associated with particulate inclusions and are found in type II carbon fibers.

When carbon fibers were heat-treated in the range 1500–3000°C and the same areas were examined using high-voltage electron micro-

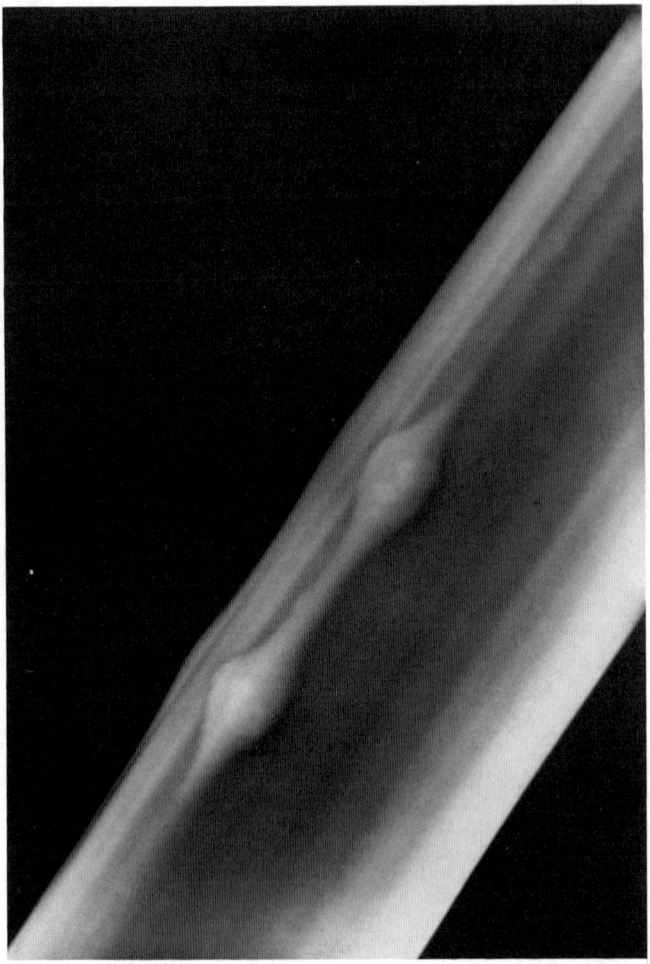

Figure 4 Diconical void. (From Ref. 68; photo courtesy of J. V. Sharp.)

scopy [68,69], the diameter of the flaw increased with increased heat treatment temperature, while the length remained almost unchanged (Figure 7). The increase in diameter differed considerably and varied from 20 to 80% over the same temperature range [68]. In addition, the inclusions originally visible at 1500°C were considerably modified. There was coalescence of distinctly separate inclusions, which then shrank. This modification of inclusions was maximum above 2000°C.

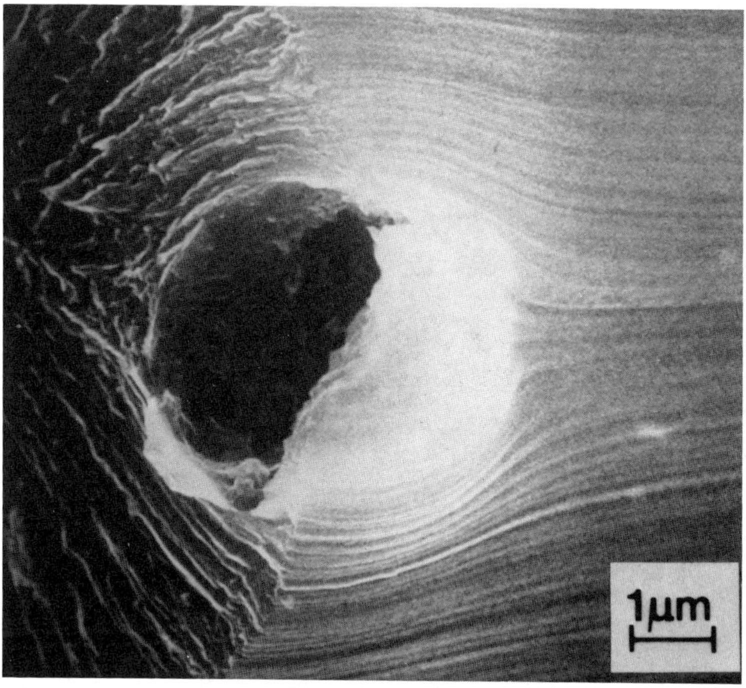

Figure 5 Inclusion. (From Ref. 68; photo courtesy of J. V. Sharp.)

4.4.2 Influence of Flaws on Strength

It is apparent that most of these strength-limiting flaws in carbon fibers are due to contamination by impurities of the precursor fiber, which react during carbonization and heat treatment to form voids and inclusions. Contamination of the fiber can occur at any stage of the carbon fiber process. Thus, particulates may be present in the spinning solution or the impurities may be picked up during spinning or further handling of the fiber. Ordinary laboratory air contains about 10^7 particles m^3 above 0.5 μm in size. Moreton and Watt [71], therefore, prepared carbon fibers in clean room conditions where the atmosphere particle content was less than 3.5×10^3 m^3. The carbon fibers were also spun under ordinary laboratory conditions for comparison. The fibers were prepared before and after filtering the spinning solution under both clean-room and ordinary laboratory conditions in order to eliminate impurities from the spinning solution. Examination of the different samples of fibers under optical microscope showed that the number of impurity particles, usually 1–3 μm

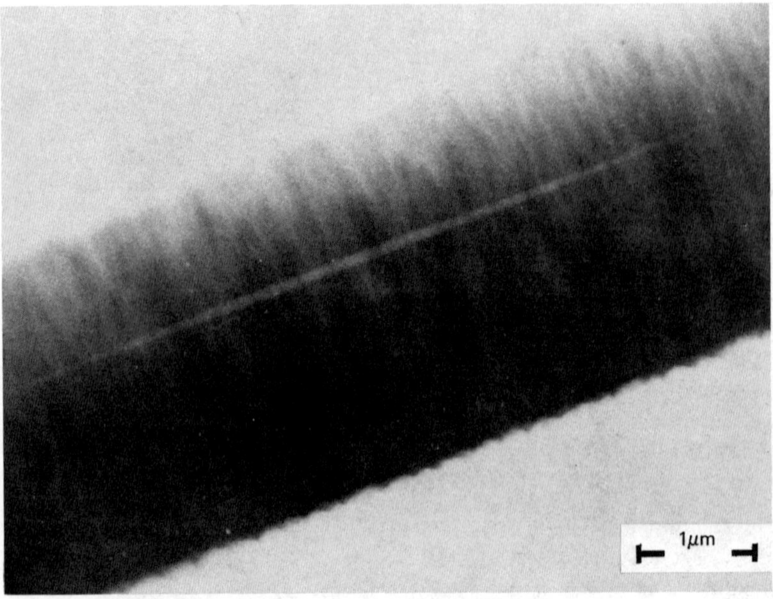

Figure 6 Needle-shaped cavity. (From Ref. 68; photo courtesy of J. V. Sharp.)

in diameter, decreased from 7 cm^{-1} to 2 cm^{-1} along 10 cm of fiber when prepared in clean-room conditions.

Carbon fibers prepared under clean-room conditions showed considerably higher strengths compared to carbon fibers obtained under ordinary laboratory conditions (Figure 8). Clean PAN fibers gave carbon fibers with a strength of 2.75 GPa after heat treatment at 2500°C, while the control fibers prepared under ordinary laboratory conditions had a strength of only 1.5 GPa, so that there was an improvement in strength of about 82%. The clean room fibers carbonized to 1000°C showed an increase in strength when heat-treated to 1400°C [71], whereas the control fibers showed a decrease. The strength at the clean fibers did not show a decrease after a heat treatment of 1500°C, which has been found to be the case for most of the fibers. The unfiltered clean sample produced carbon fibers with higher strengths than the filtered normal sample after carbonization at 1000°C, but not after heat treatment to 2500°C. This indicates that the strength is largely limited by the surface flaws at heat treatments below 1500°C. The internal flaws become more important only after heat treatment to high temperatures. This was further supported by the fact that the samples prepared under clean conditions showed little

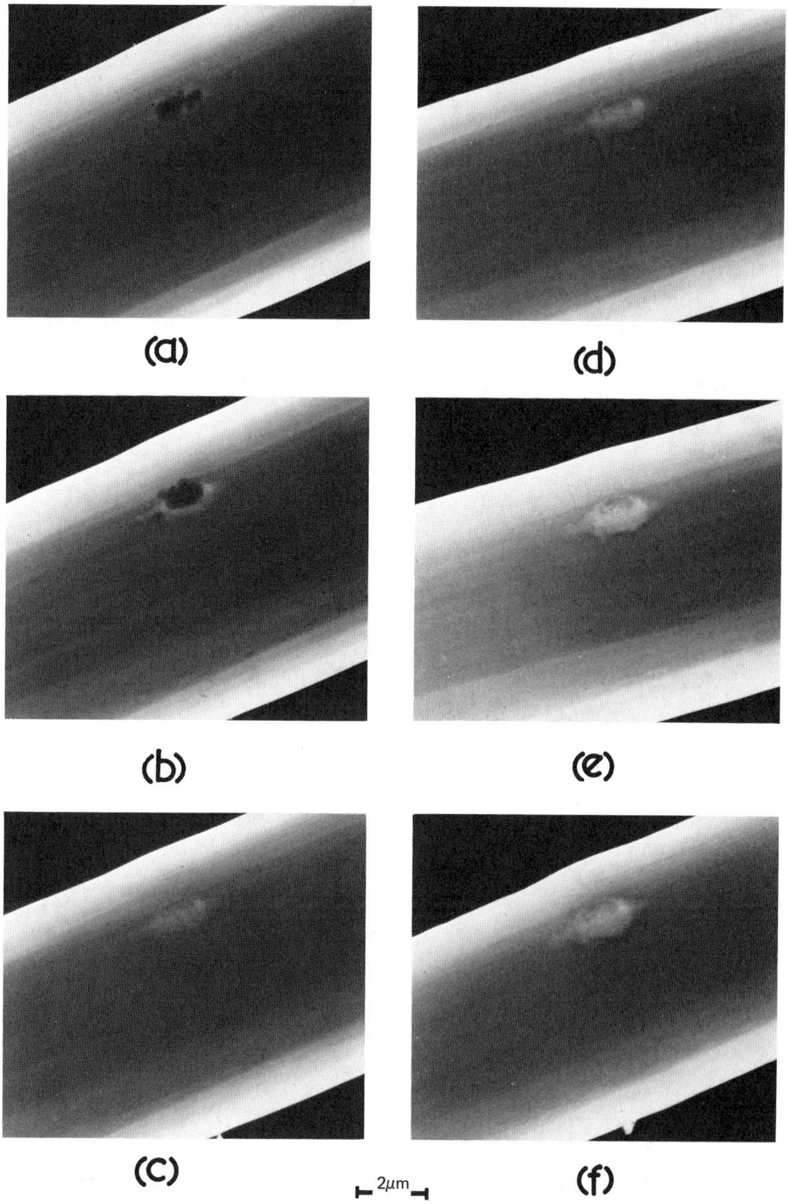

Figure 7 Development of typical internal flaw during heat treatment (a) 1500°C, (b) 1800°C, (c) 2000°C, (d) 2200°C, (e) 2620°C, (f) 3000°C. (From Ref. 68; photo courtesy of J. V. Sharp.)

Table 6 The Effect of Gauge Length on the Tensile Strength of Carbon Fibers Heat-Treated to 2500°C

Clean room fibers		
Gauge length (cm)	Average strength	Normal laboratory condition fibers
1.0	2.74	2.19
2.5	2.76	1.79
5.0	2.75	1.51

Source: Data from Ref. 71.

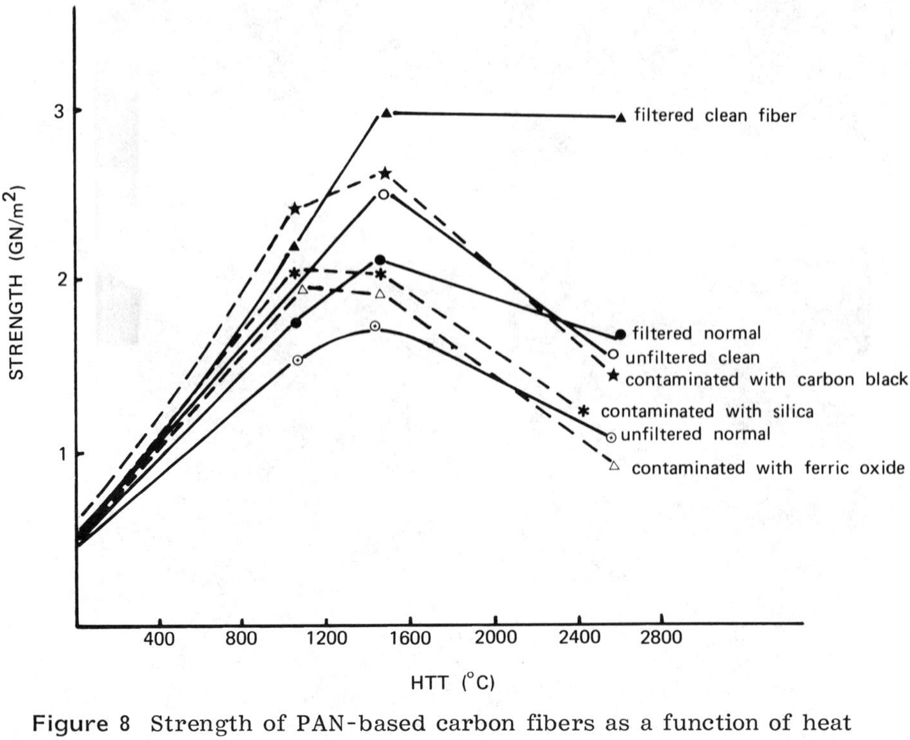

Figure 8 Strength of PAN-based carbon fibers as a function of heat treatment temperature. (From Ref. 72.)

gauge length effects compared to samples spun under normal laboratory conditions which showed marked gauge length effects (Table 6).

Reynolds and Moreton [72] deliberately contaminated PAN fibers at the final stage of the spinning process by allowing 1% suspension in water of carbon black, silica, or ferric oxide to drip onto the fibers as they were being collected over the frames. The presence of the impurities resulted in a decrease in average strength compared to the clean carbon fibers (Figure 8), and the decrease in strength in the case of the contaminated fibers became more prominent with increased HTT. The scatter in strengths was also greater in the case of contaminated fibers, the coefficient of variation being between 44–55% compared to 22% for the clean fibers [72].

The fracture surface examination of carbon fibers prepared under clean, normal, and deliberately contaminated conditions by SEM [72], showed that carbon fibers obtained from clean-room conditions and from filtered solutions generally failed at surface flaws. These flaws often appeared to be pits on one fracture surface, with a matching surface impurity particle visible on the second fracture surface. The carbon fibers prepared from unfiltered solutions failed either at surface flaws or at internal flaws when heat-treated below 1400°C, but almost invariably at cavities within the fiber after heat treatment at 2500°C. Fracture surfaces of deliberately contaminated carbon fibers showed that the surface flaws are caused by silica or iron oxide particles.

The development of these flaws from an oxide particle has been attributed [72] to a chemical reduction reaction which occurs during the carbonization and heat treatment stages. However, the variation of strength of contaminated fibers over a wide range of heat treatment temperatures indicated that chemical reaction alone was not sufficient to explain the effect, particularly in the case of ferric oxide particles, which were likely to be almost completely reduced below 1000°C. Reynolds and Moreton [72] suggested that these impurities catalyze the graphitization process [73,74] resulting in the formation of localized, randomly oriented three-dimensional graphite within the carbon fiber structure [69,6,75]. This results in an increase in the structural defects and flaws which reduce the strength with heat treatment temperature.

Some workers [68,69] are of the view that the diconical shape of the flaws in carbon fibers causes misorientation, since the basal planes tend to follow the contour of the flaws. These regions of misorientation have a high concentration of shear energy when the fiber is loaded under tension and initiates a crack. The propagation of the crack is facilitated by the enhanced crystallization around the flaw catalyzed by the impurity particles. Stereoscan pictures frequently showed a concentric lamellar graphitic structure surrounding the flaw (cavities). The thickness of the lamellar graphite sheets surrounding the cavities, in many cases, was found to be 0.2 μm, which is similar to the crack length calculated by Sharp and Burnay [69] by using the Griffith brittle fracture equation (0.1-0.3 μm) and is about one-tenth the size of the internal flaws (cavities) in their carbon fibers. This indicates that the failure must have been initiated at the cracks in the surrounding carbon. This was supported by their observation that the failure did not always take place at the largest defect in the fiber.

Whitney and Kimmel [76] estimated flaw sizes from the examination of fracture surfaces of carbon fibers heat-treated at 1500°C and used maximum flaw dimensions as the critical crack length to calculate apparent surface energy using the Griffith equation. They obtained values which were 3–10 times higher than the theoretical value, indicating that plastic deformation was taking place at the crack roots during the fracture. Le Maistre and Diefendorf [77] proposed a circumferential arrangement of the layer planes, and suggested the formation of circumferential cracks on cooling after the final heat treatment due to thermal expansion anisotropy. Cooper and Mayer [78] are of the view that the cracks are formed as a result of dislocation pileup.

The presence and role of disinclination in the fracture behavior of carbons, including carbon fiber, has been recently emphasized by White and others [128,128a].

Whatever the mechanisms of the formation of a crack may be, the lowering of strength of a carbon fiber is attributed to them. The absence of an electron microscopic easy observation of the existence of microcracks led Reynolds and Sharp [79] to propose their crystallite shear-limit model of carbon fiber strength. This model relates intrinsic strengths of all types of carbon fibers ranging from isotropic to graphite whiskers, to the strengths found in the presence of observable internal flaws. According to this model, carbon fibers would have a sufficient number of crystallites which are oriented at angles greater than the mean and are, therefore, under shear stress. These shear stresses create regions of large shear strain energy when loaded under tension. The failure occurs when the shear strain energy in a misaligned crystallite is sufficient to cause a basal-plane rupture, resulting in a crack. The cracks are possibly formed at crystallite boundaries and would propagate to reach a critical size for fiber fracture only if the local structure is sufficiently continuous and the tip of the crack always lies in a region for which the local misorientation angle exceeds a certain minimum.

For a misoriented crystallite with basal planes oriented at an angle ϕ to the fiber axis the condition for tensile failure is given by

$$\frac{1}{2} S'_{11}\sigma^2 > \frac{\gamma}{\pi c} \tag{7}$$

where γ is the surface energy in the basal plane, c the crack length and S'_{11} the effective compliance and is given by

$$S'_{11} = S_{11} \cos^4\phi + (S_{44} + 2S_{13}) \sin^2\phi \cos^2\phi + S_{33} \sin^4\phi \tag{8}$$

S_{11}, S_{44}, S_{33}, and S_{13} are crystallite elastic compliances [79]. Since the effective compliance S'_{11} increases rapidly with ϕ due to the large shear compliance factor (S_{44}), there is a greater concentration of strain energy in regions of misoriented crystallites, and consequently failure can initiate in these regions even at very small defects. When the crack grows to such a length that it can propagate through any crystallite, the final failure takes place. These workers [79] have proposed a crystal shear strain limit of 20% to provide a failure criterion for flaw-free carbon fibers and to give a general indication of the strengths likely to be found in flawed fibers.

Jones et al. [80,81] have recently carried out an analysis of flaws in high-strength mesophase carbon fibers, spun as mono- and multifilaments under clean and contaminated conditions, using SEM examination of fracture surfaces, and measured the strengths of carbon fibers as a function of the number of flaws. The carbon fibers prepared under clean conditions showed higher strengths of 3 GPa at 3.2 mm gauge length. A comparison of strengths at gauge lengths of 3.2 and 20 mm, respectively, in mono- and multifilament carbon fibers prepared under contaminated conditions, showed a marked gauge length dependence which is in accordance with the observations of Moreton [65] for PAN fibers. The strength of multifilament carbon fibers was also measured as a function of the heat treatment temperature and was found to decrease after a HTT of 2000°C in the case of contaminated carbon fibers, but remained more or less unchanged for carbon fibers prepared under clean conditions.

The major cause of failure of the contaminated carbonized filaments was particulate inclusions found either on the surface or in the core of the filament. The particles contained either iron with chromium or silicon and aluminum. Both types of these particles reacted during carbonization to form voids around the inclusion. Silicon-bearing particles were often present near the outer edges and were usually associated with localized swelling from gas. However, some of these strength-limiting flaws could not be identified on fracture surfaces of some of the strongest filaments, and it was suggested that the failure occurred either at small planes or at undetected surface pits. The distribution of different types of flaws in mono- and multifilament pitch carbon fibers and their strengths is shown in Tables 7 and 8.

These workers [81] observed a rough relationship between the size of the defect and the strength of the carbon fiber. The strength was found to vary inversely with the defect size. The strength of a carbon fiber was less than 1.4 GPa when the defect size was 3 μm or more and present anywhere in the fiber. Interior flaws between 2 and 3 μm or surface flaws between 1 and 3 μm in size limited the strength of the carbon fiber in the range 1.4–2.1 GPa. Surface and internal

Table 7 Strength and Flaw Distribution in Carbonized Monofilaments in Mesophase Pitch Carbon Fibers

			Flaw distribution (%)				
Conditions of fiber preparation	Gauge length (mm)	Average strength (GN m^2)	Mineral particles	Ferrous particles	Unknown particles	Gas bubbles	Undetected defects
Contaminated	20	1.2	nil	100			
	3.2	2.1	nil	56	11	nil	33
Clean	20	2.4	8	42	17	8	25
	3.2	3.0	10	20	nil	10	50

Source: Data from Ref. 81.

Table 8 Strength and Flaw Distribution in Carbonized Multifilaments in Mesophase Pitch Carbon Fibers

Conditions of fiber preparation	Gauge length (mm)	Average strength (GN m^2)	Flaw distribution (%)				
			Particle flaws	Small surface holes	Surface pits	Miscellaneous defects	Undetected defects
Contaminated	20	1.9	47	27	3	23	nil
	3.2	2.1	53	32	nil	15	nil
Clean	20	2.6	50	nil	17	8	25
	3.2	3.2	25	nil	38	12	25

Data after Ref. 81.

flaws between 0.7 and 2 μm in size or areas of extensive pitting resulted in failure between 2.1 and 2.8 GPa. Carbon fibers with strengths above 2.8 GPa contained particles or cavities 1 μm or less in diameter on their surface or were associated with surface pits or undetected defects. Carbon fibers which contained holes 0.5 μm or less in diameter showed strengths greater than 3.5 GPa.

4.5 INELASTIC BEHAVIOR IN CARBON FIBERS

Williams et al. [32], while studying the inelastic behavior of rayon-based carbon fibers at room temperature from bending experiments on single filaments using Sinclair loop test [31], observed that the stress-strain curves were linear up to approximately 0.5% strain, with slopes equal to the Young's modulus. However, when the stress was increased beyond 0.5%, the slope of the curve decreased, suggesting plastic yielding as occurs in metals. The curves in the case of carbon fibers were, however, reversible. A small decrease (≅10%) in initial modulus occurred upon unloading or when the test was repeated. This nonlinearity in the stress-strain behavior was attributed to local "yielding" or the breaking up of some cross-links between adjacent fibrils in order to undergo deformation independently. This deformation process involved buckling of the fibrils on the compression side of the filament. Such a plastic deformation, has been shown to take place by Bacon and Silvaggi [82]. The stress at which the "yielding" occurs in bending was approximately equal to the failure stress in a tensile test.

Jones and Johnson [83] used PAN carbon fibers heat-treated between 1000°C and 2800°C in their elastic test on an optical microscope by inducing a small loop in a fiber, trapping it under oil immersion between a cover slip and a microscopic slide, and reducing the loop size by pulling the loop ends until fracture occurred. Photographs were taken at intervals as the test proceeded. They observed that the PAN carbon fibers heat-treated to 1000°C behaved perfectly elastic to fracture and showed linear stress—strain behavior without any marked change in Young's modulus with strain. However, when the heat treatment temperature was increased, the carbon fibers tended to show nonlinear stress—strain behavior which was first detected at about 1600°C. The nonlinear behavior became more apparent as the HTT was increased beyond 2000°C. Since 2000°C is the temperature at which important structural changes in carbon fibers are known to occur, it is, therfore, most likely that this inelastic behavior of the carbon fiber is related to these structural variations. The most important structural change that occurs above this temperature is structural anisotropy which should result in a change in the mechanical properties. These workers [83] did not observe any change in section thickness throughout the bending test. Scanning electron microscopic (SEM) examination of fracture surfaces did not show any fracture occurring in a fibrillar or debonded material, as suggested by Williams et al. [32]. Their fracture surfaces, however, suggested a buckling failure beginning on the compressive side at about 0.5% strain, which intensified and spread. The ultimate failure, however, was seen to occur from the tensile side. Thus, they are of the view that the cause of the non-Hookean behavior is the structure failure in the fiber within the compressive region, under the action of high compressive forces set up in it due to bending and its own great structural anisotropy.

Hawthorne and Teghtsoonian [84] carried out elastica loop tests on a range of pitch-based and a few rayon-based carbon fibers, and studied axial compression fracture of PAN-, rayon-, and pitch-based carbon fibers by embedding short, single filaments in epoxy resin and microscopically observing the failure initiation on fiber surface during specimen compression [85]. The compression failure strains obtained from axial compression and elastica loop tests on various carbon fibers showed a similar variation with Young's modulus, indicating that the non-Hookean effects may involve the same mechanism in all types of carbon fibers. This was confirmed by the presence of microcracks in SEM micrographs of failed fiber surfaces on the compression side of both anisotropic pitch- and rayon-based fibers in the elastica loop tips. These were similar to those observed by Jones and Johnson [83] in the case of PAN carbon fibers. No evidence for fibrillar separation was found in any of the fibers examined by Hawthorne and Teghtsoonian. Thus, according to them, the axial compression strength and the non-Hookean behavior of anisotropic carbon fibers is determined by the tendency of the more ordered fiber surface to fail in compression, probably by microbuckling of highly aligned but poorly transversely coupled microfibrillar layer packets.

4.6 COMPRESSIVE STRENGTH

The compressive strength of single filaments of rayon-based carbon fibers embedded in epoxy resin was observed by Koeneman [86] to be about 0.9% of the Young's modulus. However, his measurements were based on the assumption that the fiber contraction was equal to the transverse Poisson contraction of the epoxy sample in which the fiber was embedded. Bacon, however, [1] is of the view that this assumption may not be completely correct, so that the actual fiber compressive strength should be much lower than Koeneman's values. Beasley [87] studied the influence of bond strength on the compressive strengths of unidirectional epoxy resin composites of Thornel carbon fiber and observed that beyond a certain minimum the compressive strength was not influenced by an improvement of bond strength. He observed also that the compressive strength was not significantly altered by factors such as the presence of sizing agent, the composition of the epoxy resin matrix, and the Young's modulus. The only factors which influence the compressive strength are the time and the temperature of heat treatment during processing. The compressive strength decreases both with increase in time and the temperature of heat treatment.

4.7 TORSIONAL SHEAR STRENGTH

Decrescente and Richards [34] measured the torsional shear strengths of single filaments of rayon carbon fibers and observed that all carbon fibers failed at almost the same level of shear strain (3%) [1].

4.8 DENSITY

The fiber density, which is defined as the ratio of the dry weight of a sample of carbon fiber to its geometrical volume, is an important parameter required for aerospace applications to ensure final density uniformity and to determine composite void fractions. Fluid displacement methods have been used for measuring the density of carbon fibers, but a comparison of the data is difficult because diverse fluids have been used, each of which gives a different value depending upon its size and the penetrating power. Spencer et al. [88], for example, measured the density of PAN carbon fibers heat-treated to different temperatures by displacement of helium and a silicon oil. The helium density values were always appreciably higher (Figure 9) than the silicon oil values because silicon oil could not enter some of the fine microcapillary pores owing to its larger molecular dimensions. At higher HTTs, the two values were comparable because many of the pores became closed to helium as well. These workers also observed a penetration drift in helium displacement density. The density was 1.804 g cm^3 after 17 min of helium admission and 2.040 g cm^3 after 24 hr.

Pinoli and Ambrosio [89] have developed a density gradient procedure for density measurement of carbon fibers. The yarn sample is freed of the sizing material by solvent extraction and the residual solvent and moisture removed by oven drying at 105°C with or without vacuum. The dry sample is subjected to ultrasonic agitation in benzene for about 5 min. The conditioned sample is then introduced directly into the density gradient column which is made up of benzene (density = 0.8794 g cm^3) and 1,1,2,2 tetrabromoethane (density = 2.9638 g cm^3). A timer is initiated upon introduction of the fiber

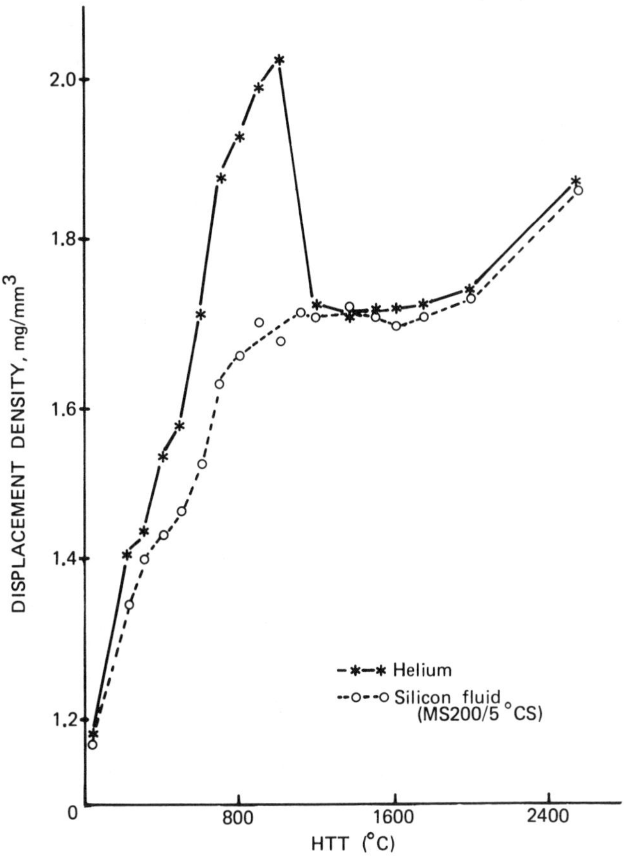

Figure 9 Displacement density of PAN-based carbon fibers at 25°C as a function of heat treatment temperature. (From Ref. 88.)

sample into the gradient column and the density values are recorded at each 5 min interval until a constant value is obtained [89].

Pinoli and Johnson [90] studied the effect of oxidation of carbon fiber and the effect of residual moisture in carbon fiber on density by the density gradient technique and observed that both factors affect the density significantly. They have, therefore, suggested care in using oxygen-free atmosphere during the high temperature heat treatment of carbon fibers and proper drying of carbon fibers before determining their density.

4.9 ELECTRONIC PROPERTIES OF CARBON FIBERS

Graphite is a moderately good conductor of electricity. It has a resistivity of 4×10^{-5} [91] in the crystallographic "a" direction, which is almost 20 times the resistivity of copper. Ubbelohde [92,95] and others [93,94] found that the resistivity of graphite can be considerably decreased by forming intercalation compounds with bromine, nitric acid, and a number of other compounds. High-modulus carbon fibers are composed of graphitic crystallites with their a axis preferentially lying parallel to the fiber axis. This orientation is most suitable in enhancing the conductivity of carbon fibers. In addition, the geometric form of the filaments is more suitable for making satisfactory contacts and for engineering design of conducting materials. Consequently a considerable amount of interest has arisen in measuring and improving the electronic properties of carbon fibers.

4.9.1 Electrical Resistivity (ρ)

The electrical resistance of a single carbon fiber is measured by stretching it under light tension on a microscopic slide. Electrical contacts are painted on the fiber and the resistance is measured by means of a standard four-probe technique using a sensitive potentiometer. Yamaguchi [96] and Robson et al. [97] studied the electrical resistivity of PAN carbon fibers as a function of HTT and observed that it decreased with increase in HTT. The decrease is more rapid below 1000°C HTT and gradual above this temperature [96]. Comparison of the resistivity with soft carbons heat-treated to the same temperatures showed it to be lower than that of soft carbons, except at high HTTs. This lower resistivity in PAN carbon fibers is due to the preferential alignment of graphitic crystallites along the fiber axis. At higher HTTs, the soft carbons have lower resistivities because of a higher degree of graphitization as compared to PAN carbon fibers. The resistivity ratio (ρ_{77} K/ρ_{300} K) in fibers decreased at lower HTTs because the resistance is more dependent on thermally activated carriers. The increase in resistivity ratio at high HTTs is due to increased graphitization. The increase in resistivity ratio commenced around 1700°C.

Rayon-based carbon fibers (WyB and WyD) showed semiconductor-type conductivities with very low activation energies of the free carriers [98]. The activation energy decreased with temperature and was nearly zero at liquid nitrogen temperature. This indicates that the Fermi level moved gradually from inside of the occupied bands at high temperature. The conductivity increased appreciably on introducing potassium into the fibers. The increase in conductivity was proportional to the number of potassium atoms introduced into the

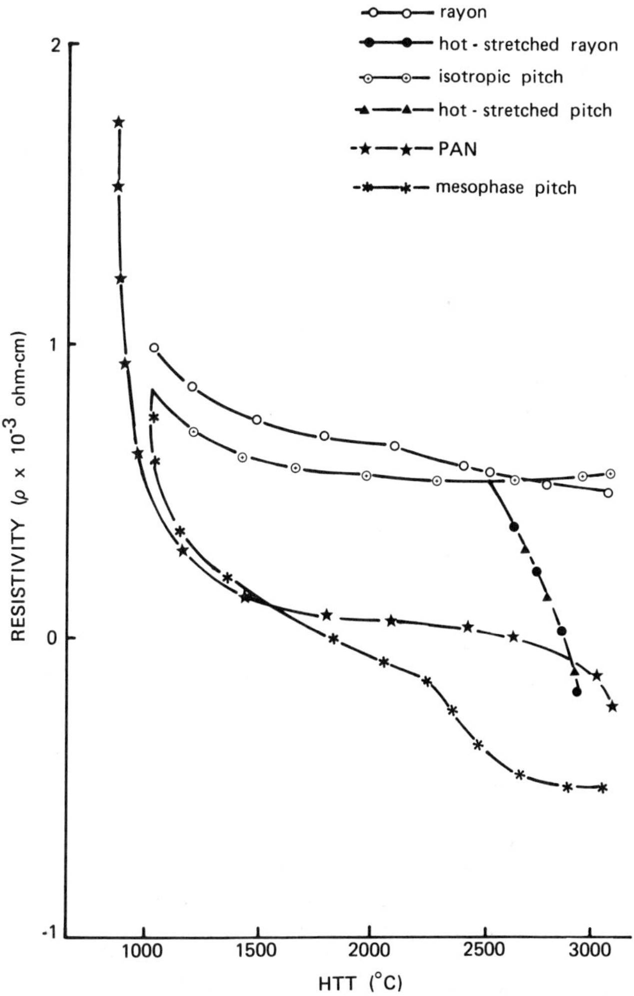

Figure 10 Resistivity of different carbon fibers as a function of heat treatment temperature. (From Ref. 99.)

fiber, each metal atom contributing one electron to the conduction band [98].

The resistivity decreased with increasing HTT in mesophase pitch carbon fibers [20] as well. At a given HTT the resistivity decreased with increasing temperature, although the precise temperature dependence was different for different HTTs. The resistivity ratio $\rho_{300\ K}/\rho_{4.2\ K}$ increased from 0.3 at 1000°C to 0.93 at 1700°C HTT and then decreased to about 0.4 at 3000°C HTT, suggesting that 1700°C is a critical temperature for inducing changes in the electronic structure of carbon fibers. The radial structure fibers showed lower resistivity compared to random structured fibers. The difference persisted even at higher temperatures and increased with increase in HTT. The resistivity of random fibers was almost double that of the radial fibers at 3000°C HTT. This is due to the fact that radial fibers are more easily graphitized than are random fibers.

Fischbach and Komaki [99[studied the electrical resistivity at 0°C of a large number of carbon fibers from PAN, rayon, isotropic pitch, and mesophase pitch as a function of HTT up to 3000°C. The resistivity decreased rapidly below 1000°C in the case of PAN carbon fibers (Figure 10). However, the rate of decrease slowed in the range of 1000–1500°C, showing a plateau, and was followed by some further decrease in the HTT range 2000–2800°C. In the case of low-modulus carbon fibers from rayon and isotropic pitch, the resistivities were similar in magnitude and remained relatively higher even after heat treatment to 3000°C (Figure 10). The resistivity of PAN carbon fibers was only one-fourth to one-fifth the value for rayon and isotropic pitch fibers at temperatures above 2000°C. In the case of mesophase pitch fibers, the resistivities were comparable to those of PAN fibers below 1600°C, but dropped considerably at high HTTs. When rayon or isotropic pitch fibers were hot-stretched the resistivities dropped to values comparable to or even lower than those for PAN carbon fibers.

Singer [130] has shown that mesophase pitch-based carbon fibers can have several transverse structures, radial, "onion-skin," random, or intermediate between these three, and that they are capable of achieving very high electrical conductivity [129].

A knowledge of the structural parameters and the influence of heat treatment and hot stretching on them in the case of different fibers indicates that the variation in resistivity is determined about equally by the degree of perfection of the graphitic layer structure and the degree of preferred orientation. Thus PAN, mesophase pitch, hot-stretched isotropic pitch, and rayon carbon fibers show smaller resistivity because they have a high degree of preferred orientation. A sharp fall in the resistivity of mesophase pitch fibers around 2300°C coincides with the commencement of layer-stacking order [99].

The variation of electrical resistance of carbon fibers with applied stress (piezoresistance) has also been investigated [100–104]. Most PAN fibers showed a linear increase in resistance above about 0.1% strain [100]. However, variable behavior was observed in the initial strain region for all but a few fibers. Berg et al. [101], however, observed an increase in resistance with tensile strain in the case of low-modulus (Whittaker type I) fibers and a decrease in resistance in the high-modulus (celanese) carbon fibers. The decrease in resistance continued until a tensile strain of 0.4%, after which it either remained constant or showed a slight increase. The piezoresistance effect was completely reversible in both high-modulus and low-modulus carbon fibers on unloading.

The resistance of as-processed carbon fibers prepared from benzene increased almost linearly and was completely reversible on unloading [102]. However, the carbon fibers, when heat-treated to 3000°C, showed nonlinearity in resistance. The resistance was zero or negative under low strain and tended to become positive at high strains. The heat-treated fibers did not return to their original value on unloading and showed a different behavior on reloading. The variation in resistance of as-prepared carbon fibers is mainly due to Poisson's contraction, while in heat-treated carbon fibers the change in resistance is caused by resistivity changes with strain, which may originate from contact resistance at grain boundaries or from the effects of stress on the electrical conduction in graphite planes [102].

Similar positive piezoresistance effect was observed in the case of as-received rayon, PAN, and mesophase pitch carbon fibers by Fischbach and Komaki [103]. The negative effect occurred, but only in a few high-modulus-grade rayon and PAN carbon fibers. Since these high-modulus fibers were found to have a large negative temperature coefficient of resistance, this negative effect was attributed to room temperature variations. Consequently, these workers studied the piezoresistance effect at constant temperature (0°C) on a large number of rayon, PAN, isotropic pitch, and mesophase pitch carbon fibers heat-treated under controlled conditions up to 2700°C [104]. The piezoresistance ($\delta R/R_0$) was positive for carbon fibers heat-treated to 1000°C from all precursors and was approximately linear. It decreased with increasing HTT and reached a limiting value only for carbon fibers with poor orientation (Figure 11). The piezoresistance effect changed sign and became negative at low strains for hot-stretched rayon, PAN (2700°C), and mesophase (2300°C) carbon fibers. This negative effect increased and became positive again at higher strains. But for mesophase pitch fibers heat-treated at temperatures higher than 2400°C, the negative effect had a very high negative value and did not become positive, even at high strains. The piezoresistance was reversible both in positive and negative ranges, but showed erratic behavior in the transition range.

Although the mechanism of the piezoresistance behavior in carbon fibers is still a matter of speculation, it has been observed [27] that all carbon fibers which showed negative piezoresistance contained a high degree of layer orientation structure and the magnitude of the negative piezoresistance effect correlated well with the development of layer size and perfection, as inferred from diamagnetic susceptibility measurements.

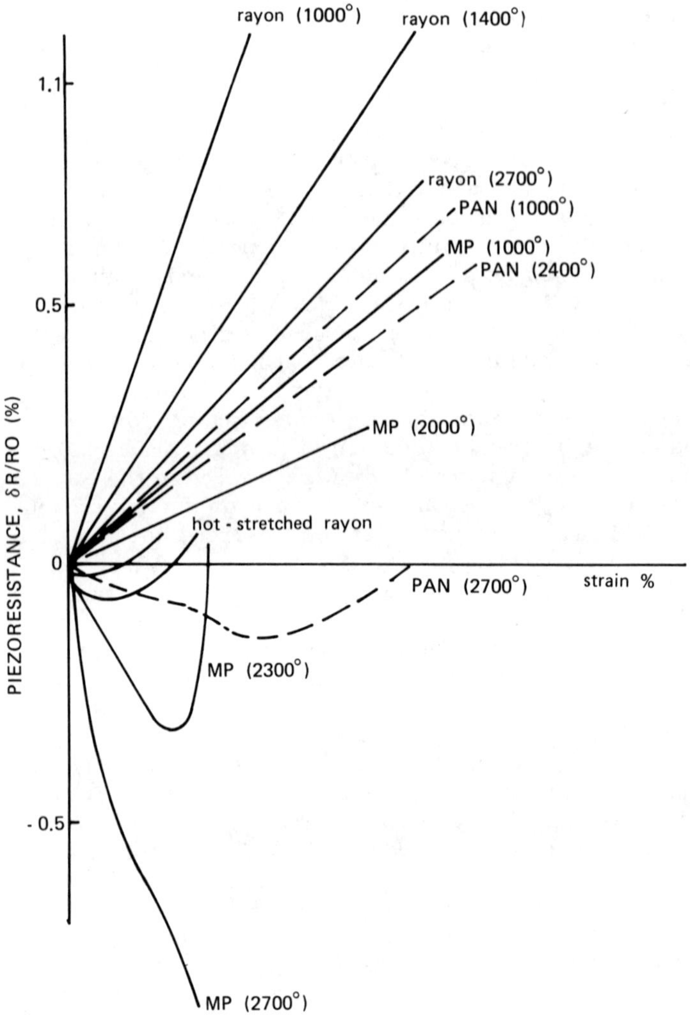

Figure 11 Piezoresistance of different carbon fibers as a function of strain. (From Ref. 104.)

4.9.2 Thermoelectric Power (TEP)

The thermoelectric power (TEP) of carbon fibers has been measured by using a thermocouple-type arrangement composed of a bundle of carbon fibers and a platinum or gold wire [96,97]. In the case of PAN fibers, the TEP decreased with increased HTT and showed a deep

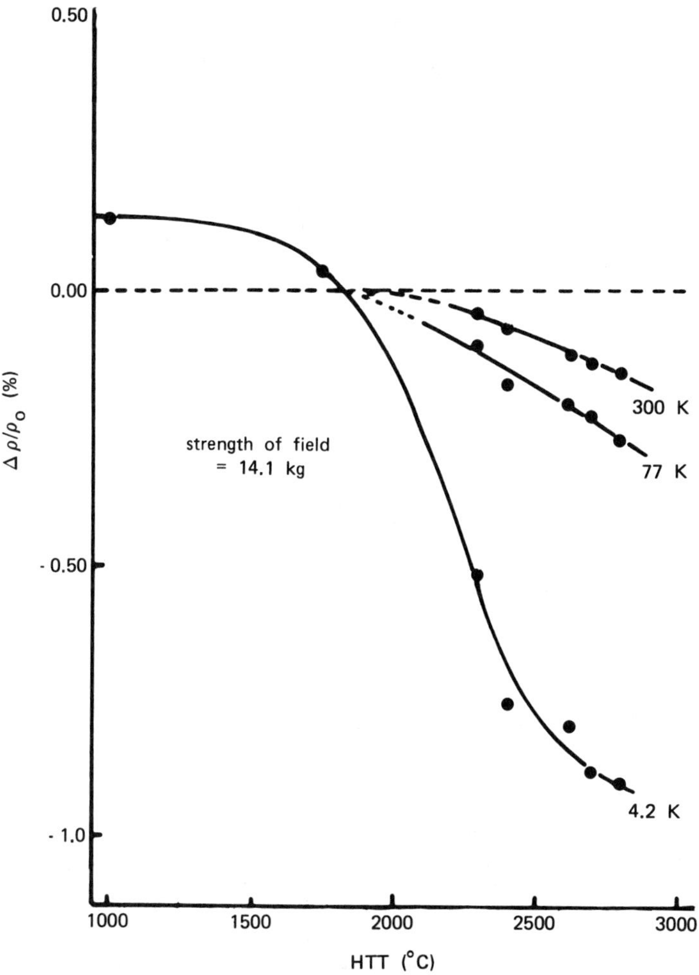

Figure 12 Transverse magnetoresistance as a function of processing temperature. (From Ref. 97.)

minimum at about 1000°C HTT, but it was always positive [66]. Robson et al. [97], on the other hand, observed that the TEP, which was negative at lower HTTs, became positive at about 1750°C, the magnitude increasing more and more as the HTT was increased. The Seebeck coefficient of rayon-based carbon fibers decreased regularly with increased HTT, but increased with the percentage of stretch [98]. All samples showed a change in the sign of thermoelectric power at low temperatures between 85 and 135 K.

4.9.3 Electron Spin Resonance

Electron spin resonaance (ESR) characteristics of a series of PAN-based carbon fibers heat-treated between 1000 and 2800°C showed an anisotropy in the g value when the HTT exceeded 1750°C. Mesophase pitch fibers showed a larger g anisotropy than PAN carbon fibers for the same heat treatment temperature [20]. Mesophase pitch fibers heat-treated in the range 2000–2300°C showed the same value of g anisotropy as PAN carbon fibers heat-treated to 3000°C. This has been attributed to the greater and easier graphitizability of mesophase pitch fibers compared to PAN fibers [20]. The characteristics of the ESR spectra were found to depend upon "single crystallite" g values, line widths, and line shapes, and upon preferred orientation of crystallites along the fiber axis, the angle between the fiber and the field, and the average extent of transverse motion by the conduction electrons [125].

4.9.4 Correlation of Various Electronic Properties

The preceding brief resume of the various electronic properties of different carbon fibers as a function of heat treatment temperature indicates that heat treatment around 1750°C is a most significant temperature for effecting changes in the electronic structure and developing of electronic properties. It is around this temperature that the anisotropy in the g value [105], appearance of negative magnetoresistance [20,97], and a change of sign of thermoelectric power [97], can be observed. A discontinuity in the x-ray parameters was also observed by Johnson and Tyson [38] in this temperature range. Furthermore, it is around this temperature that Fischbach and Komaki [104] observed a minimum in the temperature coefficient of resistivity (TCR) in the case of a large number of PAN, rayon, and pitch carbon fibers. The minimum in TCR was found to be almost independent of orientation texture or of the microstructural morphology. Fischbach and Komaki identified this range of temperature as a range in which electronic structure evolution takes place. This is consistent with the observations which have been made by several workers with other carbons.

Robson et al. [106] found that all the electronic properties of carbon fibers were related to a unique parameter defined empirically as the resistivity ratio $\rho_{77\ K}/\rho_{300\ K}$ which they called "the relative graphite order." This parameter was independent of the HTT and depended more on the actual processing conditions of the carbon fiber. They also observed a relationship between interlayer spacing and the ESR g value, indicating that the g shift arose from interlayer interactions. Thus, they are of the view that this parameter is a better representative of the electronic properties of carbon fibers as compared to HTT.

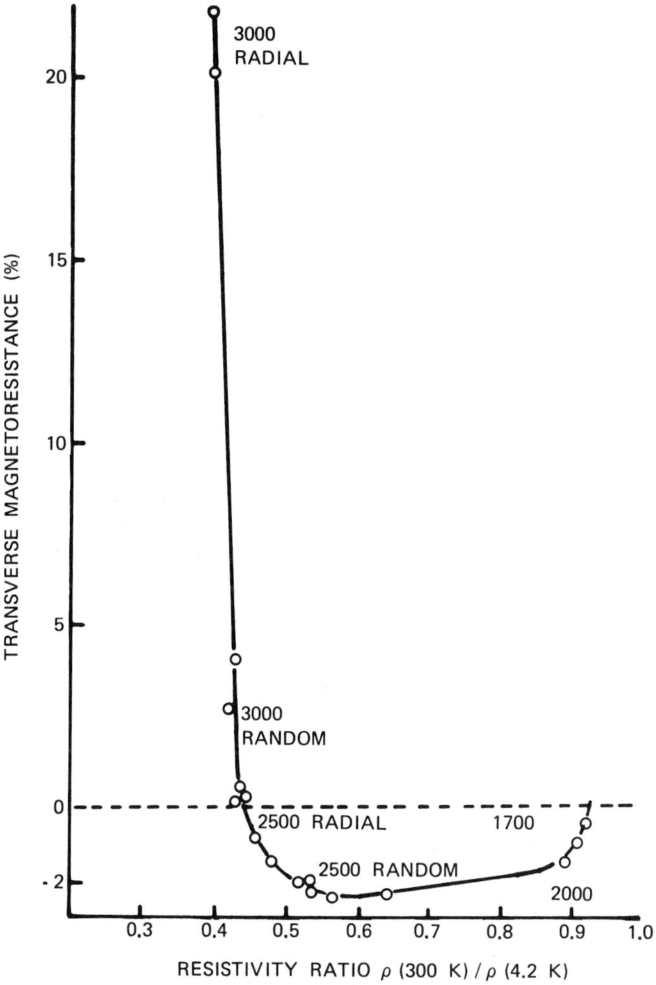

Figure 13 Transverse magnetoresistance as a function of resistivity ratio. (The temperature indicates the level of HTT.) (From Ref. 20, with permission from Pergamon Press Inc., Great Britain.)

Bright and Singer [20] also examined the relationship between the various electronic properties and the development of a graphitic character in the case of mesophase pitch carbon fibers. They observed that all the points could be accommodated on a single curve (Figure 13), the different parts of the curve corresponding to different degrees of graphitization and, therefore, only approximately to the HTT. They

[20] suggested that with increases in HTT up to 1700°C the semiconductivity energy gap in the low temperature material narrows so that the material behaves like a semimetal and the magnetoresistance tends to become negative. As the HTT is increased to 2500°C, the magnitude of the magnetoresistance increases while the resistivity ratio decreases as a result of the changes in the temperature dependence of the carrier concentration in the semimetal. Three-dimensional ordering of the graphite planes, which commences beyond 2500°C, results in the development of a graphitic type of band structure with a large positive magnetoresistance.

4.9.5 Electronic Properties of Intercalated Carbon Fibers

The resistivity of high-modulus carbon fibers from several sources after intercalation with red fuming nitric acid was examined by Vogel [107,108] with Popowich [109], both of whom observed a decrease in resistivity by an order of magnitude on intercalation. In the case of carbon fibers heat-treated to high temperatures, the changes occurred more readily with immersion in solution than by exposure to vapor. Herinckx [98] and others [110] also observed an increase in conductivity on intercalation of rayon fibers with potassium atoms. The conductivity increased by a factor of 20, but the temperature dependence of conductivity between room temperature and liquid helium temperature was similar before and after intercalation, even when the carbon fiber was saturated with potassium atoms. The increase in conductivity was dependent on the amount of potassium atoms intercalated.

The carbon fiber nitrate, obtained by treatment of carbon fibers from benzene pyrolysis with nitric acid, also showed a decrease in resistivity by about 7% of the original material after immersion of the fiber in nitric acid for 10 min [111]. The resistivity decreased more gradually with increased time of immersion of the fiber and reached a constant value after 24 hr. This is probably due to the formation of stage II intercalation compounds. The resistivity increased gradually when the carbon fiber was taken out of the nitric acid bath and kept in air until it attained a stable value; after several hours it reached a value of 15% of the initial fiber. The resistivity increased with increased temperatures, the increase being very significant above 300 K (probably due to the decomposition of the intercalation compound).

The magnetoresistance effect measured by applying the magnetic field perpendicular to the fiber axis was positive throughout the temperature range (4.2–300 K) studied. The magnetoresistance anisotropy, measured by changing the angle between the fiber axis and the magnetic field, was also retained. This suggests the existence of graphitic layers unaffected by nitrate. The thermoelectric power changed to a positive value after nitration and increased monotonically with increasing temperature. The phonon-drag effect observed in the heat-treated carbon fiber disappeared on nitration, presumably because of a disturbance of the crystal lattice and shortening of the mean free path [111].

4.10 DIAMAGNETIC SUSCEPTIBILITY OF CARBON FIBERS

The diamagnetic susceptibility of commercially available rayon-, PAN-, and pitch-based carbon fibers, as a function of HTT and structural parameters was studied by Scott and Fischbach [112] and Mellinger and Fischbach [113]. The magnetic susceptibility was measured at room temperature in air by the Faraday method using a semi-micro-analytical balance, a torsionally rigid sample suspension, and an electromagnet mounted on a rotating base with adjustable height. Bundles of carbon fibers about 1 cm long and 2 mm in diameter tied with cotton thread were used. The magnetic susceptibility was measured both in axial ($X_{\parallel}$) and in radial ($X_{\perp}$) directions, relative to the fiber axis by rotating the magnet around the sample suspended at a fixed orientation, and by noting the minimum and maximum weight changes. From this data the total (tensor trace) susceptibility $X_T = X_{\parallel} + 2X_{\perp}$) and the susceptibility anisotropy ratio $X_{\perp}/X_{\parallel}$ were calculated where $X_{\perp}$ is measured with the magnetic field perpendicular to the fiber axis and $X_{\parallel}$ with the magnetic field parallel to the fiber axis.

The total susceptibility X_T increased monotonically with increasing Young's modulus in the case of low-modulus rayon and pitch carbon fibers and approached a saturation value of 19×10^6 emu for the high-modulus rayon carbon fibers and about 19.7×10^6 emu for the highest susceptibility PAN carbon fibers [112]. The total susceptibility increased with increase in the layer dimension L_a, which in general is in agreement with the behavior of other nongraphitizing carbons [114]. The susceptibility anisotropy ratio increased linearly with Young's modulus, regardless of the type of fiber.

Variation of total magnetic susceptibility with HTT showed a weak diamagnetism for PAN carbon fibers heat-treated below 1400°C ($X_T \leqslant 1$), but it increased rapidly between 1400 and 1700°C ($X_T = 6.5$) (Figure 14). The rate of increase slowed above 2500°C and tended to approach the single crystal value (22×10^6 emu) only at very high HTTs. The anisotropy ratio was quite insignificant below 1400°C, but increased above 1400°C with a positive curvature up to 2500°C. The axial diamagnetic susceptibility $X_{\parallel}$, however, passed through a maximum near 2400°C HTT.

The very low diamagnetism for low HTT PAN carbon fibers is due to the presence of impurities, which lower the Fermi level and contribute a significant paramagnetic component, and to the evolution of nitrogen from PAN at these temperatures. The total diamagnetic susceptibility was about the same for all PAN fibers heat-treated beyond 1700°C (Table 9), irrespective of the source. This indicates that the

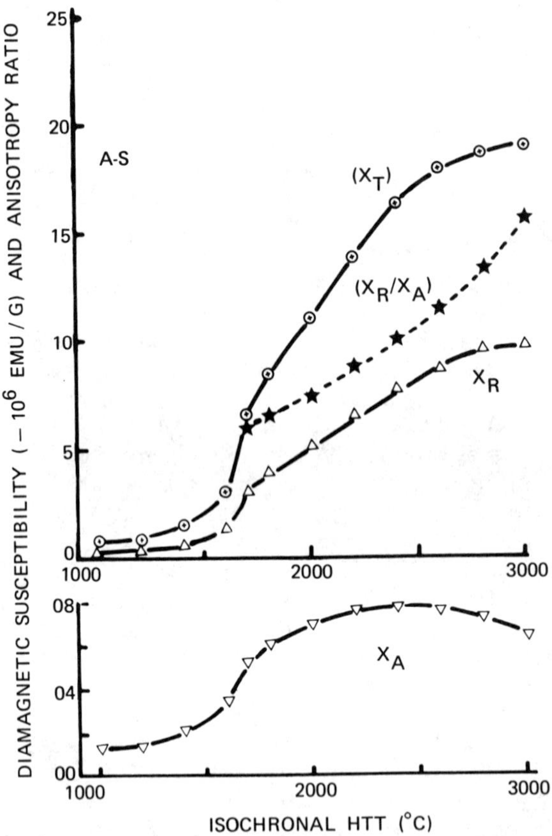

Figure 14 Diamagnetic susceptibility parameters for PAN carbon fibers as a function of heat treatment temperature. X_A = Axial diamagnetic susceptibility; X_R = radial diamagnetic susceptibility; X_T = total diamagnetic susceptibility; X_R/X_A = anisotropy ratio. (From Ref. 113.)

graphitization behavior, which determines the diamagnetic susceptibility, is independent of the variations in precursor, pretreatment, or processing history as occur in commercial practice. However, the anisotropy ratio varied significantly (Table 9) among the heat-treated fibers from PAN, probably due to differences in precursor or processing conditions. The total diamagnetic susceptibility increased with increased HTT for rayon and pitch carbon fibers as well, but the values were always lower than for PAN fibers, structure possibly playing a role, particularly for pitch-based carbon fibers.

Table 9 Diamagnetic Susceptibility Parameters for Different Carbon Fibers at 3000°C HTT

Carbon fiber	Total diamagnetic susceptibility $X_{T'}$ $(.10^{-6}$ emu/g)	Anisotropy ratio $X_{\perp}/X_{\parallel}$
A-S (PAN)	19.9	15.3
HT-U (PAN)	20.3	19.4
HT-S (PAN)	20.0	16.8
PCP (PAN)	18.6	12.0
T-400 (PAN)	20.6	23.5
VyB- (Rayon)	14.0	1.6
KCF-100 (Pitch)	13.0	1.2

Source: Data from Ref. 113.

4.11 MAGNETORESISTANCE OF CARBON FIBERS

The magnetoresistance is usually measured by using a four-probe method as in the case of the electrical resistance. The measurements can be carried out on single fibers or on a bundle of fibers. The latter is, however, preferred because it is easier to handle. The fibers are placed inside a cryostat [97] which can be placed between the poles of an electromagnet. The fiber axes are kept in a particular direction, so that when the magnet is rotated the variation in the magnetoresistance with the angle between the magnetic field and the fiber axes can be investigated. The crystal is filled with liquid nitrogen or liquid helium for low-temperature measurements.

The magnetoresistance of PAN carbon fibers at 77 and 300 K, respectively, was always negative [96]. Its magnitude depended upon tne type of carbon fiber, and decreased with the strength of the magnetic field. The magnetoresistance varied exponentially as the field strength at all values of the field and at all temperatures [97], although the value of the exponent was different for different temperatures and also depended on the sign of the magnetoresistance. The magnetoresistance was positive at lower HTTs but changed sign and became negative at about 1800°C, after which it increased in magnitude with increasing HTT (Figure 12). The effect is, however, greatly reduced when measurements are made at higher temperatures of 77 and 300 K (Figure 12).

Bright and Singer [20] investigated the longitudinal and transverse magnetoresistance of mesophase pitch carbon fibers as a function of HTT at 4.2 K. The transverse magnetoresistance was small and positive at 1000°C HTT. However, it became negative and increased in magnitude as the HTT was raised to about 2500°C. Above this HTT it became positive at high fields, and by 3000°C the magnetoresistance was positive even at low fields. The longitudinal component was immeasurably small at 1000°C HTT, but gradually grew negative with an increase in HTT. It was only slightly positive at high fields at 3000°C HTT. The radial fibers showed a larger positive magnetoresistance at 3000°C HTT than the random fibers, indicating a greater degree of graphitic order.

No magnetoresistance was observed in the case of rayon carbon fibers heat-treated to high temperatures (WyB and WyD) at liquid nitrogen temperature [98]. This was attributed to the existence of a large number of structural defects in these fibers which govern the main scattering mechanism.

4.12 THERMOPHYSICAL PROPERTIES OF CARBON FIBERS

The thermophysical properties of carbon fibers are not well understood, as very few studies have been performed on them. The measurements on single fibers are difficult and involve several problems. The results obtained show wide variations, and therefore cannot fit a unified theoretical model. We shall discuss two thermophysical properties; the coefficient of thermal expansion and thermal conductivity, on which some work is reported in the literature.

4.12.1 Coefficient of Thermal Expansion (CTE)

A carbon fiber has two principal coefficients of thermal expansion (CTE); coefficient of expansion perpendicular to the fiber axis (transverse coefficient of expansion, $\alpha_\perp$), and the coefficient of expansion parallel to the fiber axis (axial coefficient of expansion, $\alpha_\parallel$). These two coefficients have been defined by Reynolds [25] in terms of the crystallite value α_c and α_a as

$$\alpha_\perp = \left(\frac{A \cdot I_3}{2I_1}\right)\alpha_c + \left(1 - \frac{A \cdot I_3}{2I_1}\right)\alpha_a \tag{9}$$

and

$$\alpha_\parallel = A\left(1 - \frac{I_3}{I_1}\right)\alpha_c + \left(1 - A + \frac{A \cdot I_3}{I_1}\right)\alpha_a \tag{10}$$

where I_3/I_1 is the anisotropy ratio defined as

$$\frac{I_3}{I_1} = \frac{\int_0^{\pi/2} I(\phi)\, \sin^3\phi\, d\phi}{\int_0^{\pi/2} I(\phi)\, \sin\phi\, d\phi} \qquad (11)$$

and A is a parameter which takes into account the internal stresses and an accommodation coefficient due to porosity. The anisotropy ratio was found to be 0.9882, and if A is taken in the range 0.5–0.7 as in the case of graphite then

$$\alpha_{\perp} = \sim 5.5 \text{ to } 8.4 \times 10^{-6}\ \text{k}^{-1}$$

and

$$\alpha_{\parallel} = -\ 1.2 \text{ to } -\ 1.3 \times 10^{-6}\ \text{k}^{-1}$$

Fenning and Fleck [115] measured the transverse coefficient of thermal expansion ($\alpha_{\perp}$) of binderless Thornel 50 carbon fibers, bonded in tantalum tubes, using a quartz dilatometer, and obtained a value varying between 5 to 18 × 10^{-6} k^{-1} for the temperature range 100–1000°C. They compared their results with similar data obtained by Nelson and Riley [116] for single-crystal graphite and with the data for bulk graphite of various grades. The data for carbon fibers was intermediate between single-crystal and bulk graphite. The values tended to approach bulk graphite at lower temperatures and single-crystal at higher temperatures.

The axial coefficient of expansion of low- and high-modulus carbon fibers was measured by Bacon and Williams [117]. For high-modulus carbon fibers (Thornel) the value was negative below 400°C, zero at 400°C, and positive above 400°C. The coefficient of expansion was slightly larger than the single-crystal value beyond 500°C. In the case of low-modulus carbon fibers (WyB) the coefficient of expansion was positive and appreciably larger than the single-crystal value at all temperatures. Butler et al. [118] measured the axial thermal expansion of a large number of commercially available carbon fibers in the temperature range 25–2500°C, by heating carbon fiber bundles in an inert atmosphere. The length changes were measured using a linear differential transformer attached to the end of the fiber bundle. The maximum fiber bundle temperature within ±15°C was measured using a micro-optical pyrometer. The percentage expansion decreased with increases in Young's modulus of the carbon fiber, and percentage expansion versus temperature curves tended to approach that for the single-crystal "a" direction as the fiber modulus increased.

Wasan [119] has recently described a sag method for determination of the axial coefficient of the thermal expansion of carbon fibers. It in-

Table 10 Thermal Conductivities of Different Carbon Fibers

Carbon fiber	Thermal conductivity (ω cm^{-1} k^{-1})	Reference
HM (PAN)	0.594	Johnson and Watt [120]
Type I (PAN)	1.02	Knibbs et al. [121]
Type II (PAN)	0.22	Knibbs et al. [121]
Thornel 50 (RAYON)	0.90 (1800 K)	Taylor [122]
Thornel 50 (RAYON)	0.60 (room temp.)	Kalnin [123]
Morganite	0.20	Kalnin [123]

volves measuring a sag produced in a sample of carbon fiber held horizontally at the two ends and caused by its linear expansion on heating by passage of current through it. It has been calculated that an actual change of 72 μm in the length of the carbon fiber sample will produce a change in sag at 2.6 mm, which can be precisely measured using a cathetometer [119]. For Beslon-based PAN carbon fibers the axial coefficient of thermal expansion was found to be 1.0×10^{-6} k^{-1} with a standard deviation of 8×10^{-8}. More graphitizable pitch-based carbon fiber can be obtained with very low CTE [129].

4.12.2 Thermal Conductivity

The direct measurement of the thermal conductivity of carbon fibers is difficult. This value is, therefore, calculated from the measured effective thermal conductivity (K_e) of unidirectional fiber-resin composites using the equation

$$K_e = U_f \cdot K_f + U_m K_m \tag{12}$$

where U_f and U_m are the volume fractions and K_f and K_m are the thermal conductivities of the fiber and the matrix, respectively. Thus, Johnson and Watt [120], for the first time, calculated the thermal conductivity of high-modulus PAN carbon fibers from the measurements of thermal conductivity data of a composite bar at 77°C, of 0.34 volume fraction unidirectional fibers in a plastic matrix. The value of 0.594 ω $cm^{-1}k^{-1}$ compared fairly with the theoretical value of 0.71 ω cm^{-1} k^{-1} calculated by using the Debye equation. Several investigators [121–123] determined the thermal conductivities of different types of carbon fibers using the unidirectional composite method. These values are collected in Table 10. It can be seen that the results are scattered over a wide range of values. These variations are attributed to uncertainties in the measured effective thermal conductivity

Table 11 Thermal Conductivities of High-Modulus Carbon Fibers

Carbon fiber	Thermal diffusibility $\alpha(cm^2\ sec^{-1})$	Density $(g\ cm^{-1})$ $\rho(g\ cm^{-1})$	Specific heat $C(J\ g^{-1}\ u^{-1})$	Thermal conductivity $k_f\ (\omega\ cm^{-1}\ k^{-1})$ From diffusibility measurements $k_f = \alpha\ \rho C$	Thermal conductivity $k_f\ (\omega\ cm^{-1}\ k^{-1})$ From conductivity pf fiber-epoxy composite, k_e $k_f = (k_e - U_m k_m)/v_f$
Morganite II	0.690	1.82	0.937	0.118	0.136
Thornel 50S	0.365	1.66	0.962	0.583	0.592
PX-505	0.0091	1.70	0.962	0.0249	—

Source: Data from Ref. 124.

of the reinforced composite, the measured fiber volume fraction in the composite, and to the uncertainty in the equation.

Lee and Taylor [124] have determined the thermal conductivity of Morganite II, Thornel 50S, and PX-505 carbon fibers by measuring their thermal diffusivity (α) and using the equation

$$K_f = \alpha \rho C \qquad (13)$$

where ρ is the fiber density and C is the specific heat of the carbon fiber. The specific heat was measured using a differential scanning calorimeter. The thermal diffusibility was measured by the flash method, in which the front surface of a homogeneous sample was subjected to a short radiant energy pulse and the resulting temperature history of the rear surface was recorded. In the flash method a plate containing the sample holder is attached to the spring-loaded thermocouple assembly. The plate can be moved vertically or horizontally so that the thermocouple can be adjusted at any desired position on the rear surface of the sample [124]. The half-time (which is the time from the initiation of the pulse to the time when the rear face temperature rises and reaches half of its maximum) is determined at several different locations based on a hexagonal pattern. The effective thermal diffusibility (α) is then given by

$$\alpha = \frac{0.139\ \ell^2}{\Gamma} \qquad (14)$$

where ℓ is the thickness of the sample and Γ is the average value of half-times. The thermal conductivities of the carbon fibers were also calculated from the thermal conductivities of their unidirectional composites using Eq. (12). The values of thermal conductivity for Morganite II and Thornel 50S carbon fibers derived from conductivity measurements of fiber epoxy composites are within 15% and 2%, respectively, of the values determined from thermal diffusibility measurements (Table 11). These are well within the uncertainties involved in thermal conductivity measurements. The conductivity value of 0.583 $\omega\ cm^{-1}\ k^{-1}$ for Thornel 50S, calculated from diffusibility measurements, is also in good agreement with the value of 0.60 $\omega\ cm^{-1}\ k^{-1}$ reported by Kalnin et al. [123] although the value of 0.118 $\omega\ cm^{-1}\ k^{-1}$ for Morganite II differs from the value of 0.2 $\omega\ cm^{-1}\ k^{-1}$ reported by Kalnin et al. [123].

REFERENCES

1. Bacon, R. *Chemistry and Physics of Carbon*, vol. 9, (Edited by P. L. Walker, Jr. and P. Thrower) Marcel Dekker, New York, 1973, pp. 1–102.

2. Fourdeux, A., Perret, R., and Ruland, W. *Carbon Fibres—Their Composites and Applications*, The Plastics Institute, London, 1971, p. 57.
3. Ruland, W. Appl. Polymer Symposia, No. 9, 293 (1969).
4. Tokarsky, E., and Diefendorf, R. J. *12th Bienn. Conf. on Carbon*, Pittsburgh, Penn., 1975, extended abstracts, p. 301.
5. Stevens, W. C., Wang, J. H., and Diefendorf, R. J. *Carbon 80, 3rd Intern. Carbon Conf.*, Baden-Baden, 1980 preprints, p. 598.
6. Johnson, J. W., Rose, P. G., and Scott, G. *Proc. 3rd. Conf. on Ind. Carbon and Graphite*, Soc. Chem. Ind., London, 1971, p. 443.
7. Jones, B. F., and Duncan, R. G. J. Mater. Sci. *6*, 289 (1971).
8. Perry, A. J., Phillips, K., and Delamotte, E. Fibre Sci. Technol. *3*, 317 (1971).
9. Delamotte, E., and Perry, A. J. Fibre Sci. Technol. *3*, 157 (1971).
10. Bacon, R., and Schalamon, W. A. 8th Bienn. Conf. on Carbon Buffalo, N.Y., 1967.
11. Gibson, D. W., and Langlois, G. B. Polymer Prepr. *9*, 1376 (1968).
12. Watt, W., and Johnson, W. *Proc. 3rd Conf. on Ind. Carbon and Graphite*, Soc. Chem. Ind., London, 1971, p. 417.
13. Bahl, O. P., and Manocha, L. M. Carbon *13*, 297 (1975).
14. Watt, W., and Johnson, W. Appl. Polymer Symposia No. 9, 215 (1969).
15. Bahl, O. P., and Manocha, L. M. *12th Bienn. Conf. on Carbon*, Pittsburgh, Penn., 1975, extended abstracts, p. 323.
16. Moreton, R. *Proc. 3rd Conf. on Ind. Carbon and Graphite*, Soc. Chem. Ind., London, 1971, p. 472.
17. Baker, C., Bentall, R. H., Hawthorne, H. M., and Singer, K. R., *Proc. 3rd Conf. on Ind. Carbon and Graphite*, Soc. Chem. Ind., London, 1971, p. 508.
18. Otani, S., and Yokoyama, A. Appl. Polymer Symposia No. 9, 325 (1969).
19. Johnson, D. J. *Carbon Fibres—Their Composites and Applications*, The Plastics Institute, London, 1971, p. 52.
20. Bright, A. A., and Singer, L. S. Carbon *17*, 59 (1979).
21. Allen, S., Cooper, G. A., Johnson, D. J., and Mayer, R. M. *Proc. 3rd Conf. on Ind. Carbon and Graphite*, Soc. Chem. Ind., London, 1971, p. 456.
22. Wicks, B. J. J. Nucl. Mater. *56*, 287 (1975).
23. Henricksen, R. E., and Fischbach, D. B. *4th London Intern. Conf. on Carbon and Graphite* 1974, Soc. Chem. Ind., London, 1976, p. 626.

24. Hawthorne, H. M. *Carbon Fibres—Their Composites and Applications*, The Plastics Institute, London, 1971, p. 81.
25. Reynolds, W. N. *Proc. 3rd Conf. on Ind. Carbon and Graphite*, Soc. Chem. Ind., London, 1971, p. 417.
26. Adams, R. D. J. Phys. D. Appl. Phys. *8*, 738 (1975).
27. Fischbach, D. B. *Carbon '80, 3rd Intern. Carbon Conf.*, Baden-Baden, 1980, preprints p. 590.
28. Fischbach, D. B., and Srinivasagopalan, S. *5th London Intern. Carbon and Graphite Conf.*, vol. 1, 1978, Soc. Chem. Ind., London, 1978, p. 389.
29. Henricksen, R. E., and Fischbach, D. B. *12th Bienn. Conf. on Carbon*, Pittsburgh, Penn., 1975, extended abstracts, p. 626.
30. Diefendorf, R. J., and Tokarsky, E. Polym. Engg. and Sci. *15*, 150 (1975).
31. Sinclair, D. J. Appl. Phys. *21*, 380 (1950).
32. Williams, W. S., Steffens, D. A., and Bacon, R. J. Appl. Phys. *41*, 4893 (1970).
33. Bacon, R. *Chemistry and Physics of Carbon*, vol. 9, (Edited by P. L. Walker, Jr. and P. Thrower) Marcel Dekker, New York, 1973, p. 1–102.
34. Decrescente, M. A., and Richards, R. Quoted in *Chemistry and Physics of Carbon*, vol. 9, (Edited by P. L. Walker, Jr. and P. Thrower), Marcel Dekker, New York, 1973, pp. 1–102.
35. Moreton, R., Watt, W., and Johnson, W. Nature *222*, 690 (1967).
36. Goodhew, P. J., Clarke, A. J., and Bailey, J. E. Mater. Sci. Engg. *17*, 3 (1975).
37. Shindo, A. Govt. Industrial Research Institute, Osaka, Report no. 317, 1961.
38. Johnson, D. J., and Tyson, C. N. J. Phys. D: Appl. Phys. *3*, 526 (1970).
39. Mrozowski, S. *Proceedings 1st Conf. on Carbon*, (Edited by S. Mrozowski) Waverly Press, New York, 1956, p. 31.
40. Sharp, J. V., and Burnay, S. G. *Carbon Fibres—Their Composites and Applications*, The Plastics Institute, London, 1971, p. 68.
41. Le Maistre, C. W. Ph.D. thesis, Rennsselaer Polytechnic, New York, 1972.
42. Moreton, R. Personal Comm., quoted in Ref. 36.
43. Johnson, J. W., and Thorne, D. J. Carbon 7, 659 (1969).
44. Johnson, J. W. Appl. Polymer Symposia No. 9, 229 (1969).
45. Thorne, D. J. Nature *225*, 1039 (1970).
46. Bahl, O. P., and Manocha, L. M. *12th Bienn. Conf. on Carbon*, Pittsburgh, Penn., 1975, extended abstracts, p. 323.

47. Fitzer, E., and Müller Th. *15th Bienn. Conf. on Carbon*, Philadelphia, Penn., 1981, extended abstracts, p. 312.
48. Fitzer, E., and Müller, Th. *15th Bienn. Conf. on Carbon*, Philadelphia, Penn., 1981, extended abstracts, p. 314.
49. Johnson, W. *Proc. 3rd Conf. on Ind. Carbon and Graphite*, Soc. Chem. Ind., London, 1971, p. 447.
50. Bahl, O. P., Mathur, R. B., Seth, R. L., and Dhami, T. L. *Carbon '80 Third Intern. Carbon Conf.*, Baden-Baden, 1980, summary of papers, p. 602.
51. Kawamura, K., and Jenkins, G. M. J. Mater. Sci. *5*, 262 (1970).
52. Hsiung, P. L., and Rowe, C. R. *14th Bienn. Conf. on Carbon*, Pennsylvania State University, University Park, Penn., 1979, extended abstracts, p. 226.
53. Cooper, G. A., and Mayer, R. M. J. Mater. Sci. *6*, 60 (1971).
54. Jones, B. F., and Peggs, I. D. Nature *239*, 95 (1971).
55. Jones, B. F., and Peggs, I. D. J. Nucl. Mater. *40*, 141 (1971).
56. Reynolds, W. N. *Chemistry and Physics of Carbon*, vol. 11, (Edited by P. L. Walker, Jr. and P. Thrower), Marcel Dekker, New York, 1973, p. 1.
57. Jackson, P. W., and Marjoram, J. R. Nature *218*, 83 (1968).
58. Jackson, P. W., and Marjoram, J. R. J. Mater. Sci. *5*, 9 (1970).
59. Braddick, D. M., Jackson, P. W., and Walker, P. J. J. Mater. Sci. *6*, 419 (1971).
60. Barclay, R. B., and Borfield, W. J. Mater. Sci. *6*, 1076 (1971).
61. Johnson, J. W. Polymer Reprints, Amer. Chem. Soc. Atlantic City Meeting of Division of Polymer Chemistry, 1968, *9*(2), 1316.
62. Watt, W. Proc. Roy. Soc. London *A319*, 5 (1970).
63. Thorne, D. J. J. Appl. Polymer Sci. *14*, 103 (1970).
64. Johnson, J. W., and Thorne, D. J. Carbon *7*, 659 (1969).
65. Moreton, R. Fibre Sci. Technol. *1*, 273 (1968).
66. Chwastiak, S., Barr, J. B., and Didchenko, R. Carbon *17*, 49 (1979).
67. Thorne, D. J. Nature *248*, 754 (1974).
68. Sharp, J. V., Burnay, S. G., Matthews, J. R., and Harper, E. A. *Carbon Fibres—Their Place in Modern Technology*, The Plastics Institute, London, 1974, p. 25.
69. Sharp, J. V., and Burnay, S. G. *Carbon Fibres—Their Composites and Applications*, The Plastics Institute, London, 1971, p. 68.
70. Jarro, M. A. A., Ladner, W. R., and Rantell, T. D. Carbon *14*, 219 (1976).

71. Moreton, R., and Watt, W. Carbon *12*, 543 (1974).
72. Reynolds, W. N., and Moreton, R. Phil. Trans. Roy. Soc. London *A294*, 451 (1980).
73. Ischikawa, T., and Yochizawa, S. J. Chem. Soc. Japan *66*, 933 (1963).
74. Baranieski, C., Pinchbeck, P. H., and Pickering, F. B. Carbon *7*, 213 (1969).
75. Johnson, D. J., and Tomizuka, I. *Carbon Fibers — Their Place in Modern Technology*, The Plastics Institute, London, 1974, p. 20.
76. Whitney, E., and Kimmel, R. M. Nature Phys. Sci. *237*, 93 (1972).
77. Le Maistre, C. W., and Diefendorf, R. J. *10th Bienn. Conf. on Carbon*, Lehigh University, Bthlehem, Penn., 1971, summary of papers, p. 163.
78. Cooper, G. A., and Mayer, R. M. J. Mater. Sci. *6*, 60 (1971).
79. Reynolds, W. N., and Sharp, J. V. Carbon *12*, 103 (1974).
80. Jones, J. B., Barr, J. B., and Smith, R. E. *14th Bienn. Conf. on Carbon*, Pennsylvania State University, University Park, Penn., 1979, extended abstracts, p. 22.
81. Jones, J. B., Barr, J. B., and Smith, R. E. J. Mater. Sci. *15*, 2455 (1980).
82. Bacon, R., and Silvaggi, A. F. Carbon *6*, 231 (1968).
83. Jones, W. R., and Johnson, J. W. Carbon *9*, 645 (1971).
84. Hawthorne, H. M., and Teghtsoonian, E. *12th Bienn. Conf. on Carbon*, Pittsburgh, Penn., 1975, extended abstracts, p. 295.
85. Hawthorne, H. M., and Teghtsoonian, E. J. Mater. Sci. *10*, 41 (1975).
86. Koeneman, J. B., Ph. D. thesis, Case Western Reserve University, Cleveland, Ohio, 1970.
87. Beasley, W. C. Paper presented at the ASTM National Conference on High Performance Fibres— Properties, Applications and Test Methods, Williamsburg, Va., Nov. 17–18, 1970.
88. Spencer, D. H. T., Hooker, M. A., Thomas, A. C., and Napier, B. A. *Proc. 3rd Conf. on Ind. Carbon and Graphite*, Soc. Chem. Ind., London, 1971, p. 467.
89. Pinoli, P. C., and Ambrosio, A. *13th Bienn. Conf. on Carbon*, Irvine, Calif. 1977, extended abstracts, p. 292.
90. Pinoli, P. C., and Johnson, A. C. *13th Bienn. Conf. on Carbon*, Irvine, Calif., 1977, extended abstracts, p. 294.
91. Spain, I. L., Ubbelohde, A. R., and Young, D. A. *Proc. 2nd Intern. Carbon and Graphite Conf.* 1967.
92. Ubbelohde, A. R. Proc. Roy. Soc. London *A321*, 445 (1971).
93. Bach, B., and Ubbelohde, A. R. Proc. Roy. Soc. London *A325*, 437 (1971).

94. Murray, J. J., and Ubbelohde, A. R. Proc. Roy. Soc. London *A312*, 371 (1969).
95. Ubbelohde, A. R. Proc. Roy. Soc. London *A327*. 290 (1971).
96. Yamaguchi, T. Carbon *2*, 95 (1964).
97. Robson, D., Assabghy, F. Y. I., and Ingram, D. J. E. J. Phys. D Appl. Phys. *5*, 169 (1972).
98. Herinckx, C. *9th Bienn. Conf. on Carbon*, Boston, Mass., 1969, summary of papers, p. 46.
99. Fischbach, D. B., and Komaki, K. *14th Bienn. Conf. on Carbon*, Pennsylvania State University, University Park, Penn., 1979, extended abstracts, p. 191.
100. Owston, C. N. J. Phys. D: Appl. Phys. *3*, 1615 (1970).
101. Berg, C. A., Cumpston, H., and Rinsky, A. Text. Res. J. *42*, 486 (1972).
102. Morinobu, E., and Tsuneo, K. *14th Bienn. Conf. on Carbon*, Pennsylvania State University, University Park, Penn., 1979, extended abstracts, p. 258.
103. Fischbach, D. B., and Komaki, K. *13th Bienn. Conf. on Carbon*, Irvine, Calif. 1977, extended abstracts, p. 298.
104. Komaki, K., and Fischbach, D. B. *14th Bienn. Conf. on Carbon*, Pennsylvania State University, University Park, Penn., extended abstracts, p. 193.
105. Robson, D., Assabghy, F. Y. I., and Ingram, D. J. E. J. Phys. D: Appl. Phys. *4*, 1426 (1971).
106. Robson, D., Assabghy, F. Y. I., Cooper, E. G., and Ingram, D. J. E. J. Phys. D: Appl. Phys. *6*, 1822 (1973).
107. Vogel, F. L. *Proc. 4th London Intern. Carbon and Graphite Conf. 1974*, Soc. Chem. Ind., London, 1976, p. 332.
108. Vogel, F. L. Carbon *14*, 175 (1976).
109. Vogel, F. L., and Popowich, R. *Petroleum Derived Carbons*, Am. Chem. Soc., Symposium Series No. 21, Am. Chem. Soc., 1975, p. 411.
110. Herinckx, C., Perret, R., and Ruland, W. Nature *220*, 63 (1968).
111. Endo, M., Koyama, T., and Inagaki, M. *13th Bienn. Conf. on Carbon*, Irvine, Calif., 1977, extended abstracts, p. 299.
112. Scott, C. B., and Fischbach, D. B., *10th Bienn. Conf. on Carbon*, Lehigh University, Bethlehem, Penn., 1971, summary of papers, p. 321.
113. Mellinger, G. B., and Fischbach, D. B. *12th Bienn. Conf. on Carbon*, Pittsburgh, Penn., 1975, extended abstracts, p. 277.
114. Fischbach, D. B. Carbon *5*, 565 (1967).
115. Fenning, R. C., and Fleck, J. N. *10th Bienn. Conf. on Carbon*, Lehigh University, Bethlehem, Penn., 1971, summary of papers, p. 47.
116. Nelson, J. B., and Riley, D. P. Phys. Soc. 57, 472 (1945).

117. Bacon, R., and Williams, W. S. Unpublished Tech. Report, 1967 quoted in *Chemistry and Physics of Carbon*, vol. 9, (Edited by P. L. Walker, Jr. and P. Thrower), Marcel Dekker, New York, 1973, p. 1.
118. Butler, B., Duliere, S., and Tidmore, J. *10th Bienn. Conf. on Carbon*, Lehigh University, Bethlehem, Penn., 1971, summary of papers, p. 45.
119. Wasan, V. P. Carbon *17*, 55 (1979).
120. Johnson, W., and Watt, W. Nature *215*, 384 (1967).
121. Knibbs, R. H., Baker, D. J., and Rhodes, G. 26th Ann. Tech. Conf., Reinforced Plastics/Composites Div. Soc. of Plastics Ind. (1971).
122. Taylor, R. E. U.S. Air Force Report AFML-TR-72-9, Air Force Materials Laboratory, 1972, p. 55.
123. Kalnin, I. L., Ram, M. J., and Dix, R. U.S. Air Force Report AFML-TR-72-151 Part I, Air Force Materials Laboratory, 1942.
124. Lee, H. J., and Taylor, R. E. Carbon *13*, 521 (1975).
125. Jones, J. B., and Singer, L. S. *15th Bienn. Conf. on Carbon*, Philadelphia, Penn., 1981, extended abstracts, p. 306.
126. Messiet, J., Raedersdorff, A., and Donnet, J. B. Carbon 80 3rd Intern. Carbon Conf., Baden-Baden, 1980, p. 209.
127. Donnet, J. B. Carbon, *20*, 266, 1982
127a. Dauksch, H. Thesis, Centre Universitaire du Haut-Rhin, 1972.
128. White, J. L. Int. Symp. on Carbon, Toyohashi, Japan, 1982, extended abstract, p. 149.
128a. Ng C. B. Henderson, Buochler, M., White, J. L., Int. Symp. on Carbon, Toyohashi, Japan, 1982, extended abstracts, p. 149.
129. Singer, L. S. Int. Sympos. on Carbon, Toyohashi, Japan, 1982, extended abstracts, p. 400.
130. Singer, L. S. Carbon, *16*, 409 (1978).

5

Carbon Fiber Applications

The more important characteristics of an ideal engineering material are its high strength, high toughness, and low weight. Conventional engineering materials, the metals and their alloys, are strong and tough but not light. Some covalent materials are strong but not very tough, whereas certain plastic materials are light but lack strength and toughness. Several types of fibrous materials have been developed during the last decade and have been successfully used for certain applications. The properties of some of these materials are compared in Table 1. High-performance carbon fibers with improved strength and stiffness and very low density are one of the more recent developments. These carbon fibers are elastic to failure at normal temperature, which renders them creep-resistant and nonsusceptible to fatigue. They are chemically inert except in strongly oxidizing environments or in contact with certain molten metals. Carbon fibers have exceptional thermophysical properties and excellent damping characteristics.

The physical properties of carbon fibers which make them versatile materials for certain applications are given in Table 2. It may, however, be mentioned that specific strength and stiffness are the basic characteristics which are useful in almost all applications.

Carbon fibers, however, have some shortcomings as well. They are brittle, have a low impact resistance and low break extension, and very small coefficient of linear expansion. They have a high degree of anisotropy both in the direction of the fiber axis and perpendicular to it. They remain expensive materials and can, therefore, be used only where cost is not a major factor. However, the price of these fibers is likely to decrease with increased production.

The two main sectors of carbon fiber applications are the high technology sector, which includes aerospace and nuclear engineering, and the general engineering and transportation sector, which includes engineering components such as bearings, gears, cams, fan blades, etc., and automobile bodies. However, the requirements of the two sectors are fundamentally different. For example, the large-scale use of carbon fibers in aircraft and aerospace is driven by maximum performance and fuel efficiency, while the cost factor and the production

Table 1 Comparison of Properties of Various Engineering Materials

Material	Specific gravity	Ultimate tensile strength (MN m2)	Young's modulus (GN m2)	Specific strength (MN m2)	Specific stiffness (GN m2)
Steel	7.8	1018	205	130.5	26.2
Titanium alloy (DTD 5173)	4.5	940	112	208.9	24.9
Aluminum alloy (L 65)	2.8	470	70	167.8	25
Magnesium alloy (DTD 88C)	1.8	280	42	155.5	23.3
Glass fiber	2.5	1725	70	690	28
Type I PAN-based carbon fiber	1.90	1600	385	1370	192.5
CFRP made from type I carbon fiber	1.5	735	154	490	102.7
CFRP made from type II carbon fiber	1.4	945	107	675	76.4
Kevlar 29	1.44	2700	130	1860	90
Boron (on tungsten)	2.50	3100	160	1240	64
Silicon carbide (on tungsten)	3.50	3500	400	1000	114

requirements are not critical. The use of carbon fibers in general engineering and surface transportation is dominated by cost constraints, high production rate requirements, and generally less critical performance needs. This leads to very different considerations regarding material forms, acceptable matrices, the significance of hybridization, potential manufacturing methods, and on-line quality control requirements.

The availability of carbon fibers has, thus, made it possible to make entirely different kinds of high-performance articles. Carbon fibers, however, are seldom used alone. They are used for reinforcing certain matrix materials to form composites.

Carbon fibers are unidirectional reinforcements and can be arranged in such a way in the composite that it is stronger in the direction which must bear loads. Another arrangement of fibers can be used to prevent deflection on loading in one direction and not in the

Table 2 Characteristics and Applications of Carbon Fibers

Physical characteristics	Applications
1. Specific strength, specific toughness, light weight	Aerospace, road and marine transport, sports goods
2. High-dimensional stability, low coefficient of expansion and low abrasion	Missiles, aircraft brakes aerospace antenna
3. Good vibration damping, strength and toughness	Audio equipment
4. Electrical conductivity	Automobile hoods
5. Biological inertness and x-ray permeability	Medical application in prosthesis, surgery, and x-ray equipment
6. Fatigue resistance, self-lubrication, high damping	Textile machinery and general engineering
7. Chemical inertness, high corrosion resistance	Chemical industry, nuclear field

other. But, in general, the processed bodies are expected to bear loads, resist deflection, and give high performance in more than one direction. Consequently, the carbon fibers have to be arrayed in more than one direction in the composite to attain better efficiency. This selective reinforcement by carbon fibers allows designers to fashion articles suited to their needs. The designer may arrange more fibers in one direction than in the other, or use two types of carbon fibers, one known for its high strength to bear loads in one direction and the other for its high stiffness to resist deflection in the other direction. In addition, the structure can be constructed by using two different kinds of fiber materials such as Kevlar or glass and carbon fibers. The structures constructed using Kevlar and carbon fibers simultaneously in the same composite have impact strengths better than those reinforced with carbon fibers alone and have better compressive strength than those reinforced with Kevlar

Table 3 Companies Manufacturing Carbon Fiber Cloth

1.	Fabric Development, Inc., Quakertown, PA
2.	Fibrite Corporation, Winona, MN
3.	Fiber Materials, Inc., Biddeford, ME
4.	Hitco, Gardena, CA
5.	Hexcell Corporation, Dublin, CA
6.	Hercules, Inc., Magna, UT
7.	Prodesco, Inc., Perkosie, PA
8.	Textile Design and Development, Inc., Levittown, PA

fibers alone [1]. These composites in which more than one kind of reinforcement is used are called hybrids.

Since a composite material needs several different orientations of the reinforcing fiber for most of the applications, it is better to use the fiber in fabric form. Carbon fiber, being brittle, is difficult to obtain in fabric form, but due to advances in sizing and weaving technology, it is now possible to obtain carbon fiber cloth of superior quality in good quantity. Carbon fiber cloth has certain distinct advantages in making composites. The fabric is easy to handle and can be laid up faster in the desired shapes. It is bidirectional and can, therefore, be woven with two different types of fibers in the warp and in the fill. The labor cost in making composites in complex designs is much less and can compensate for the high cost of the woven material. Companies that currently weave carbon fiber cloth are given in Table 3.

The physical properties of carbon fiber-reinforced composite materials depend considerably on the nature of the matrix, the fiber alignment, the volume fraction of the fiber and matrix, and on the molding conditions. Therefore, by choosing suitable conditions, it is possible to make composites suited to a particular need. Several types of matrix materials such as glass and ceramics [2–5], metals [6–11], and plastics [12–20] have been used as matrices for reinforcement by carbon fibers. However, from the point of view of mechanical properties, density, and fiber-matrix cohesion, epoxy resins are frequently the best choice, although these matrices are not suitable for high-temperature applications. Several high-temperature plastic matrices have now been developed which may improve the temperature range of application of these composites. For very high-temperature applications, carbon–carbon composites consisting of a fibrous carbon substrate in a carbonaceous matrix have been developed [21–32].

5.1 TYPES OF COMPOSITES

Three main types of composite materials which have been investigated and generally prepared using carbon fibers are:

(1) Carbon–polymer composites (CFRP)
(2) Carbon–metal composites
(3) Carbon–carbon composites (CFRC).

5.1.1 Carbon–Polymer Composites (CFRP)

Carbon fibers are considered excellent reinforcements for plastic materials because (1) they themselves are strong, stiff, and lightweight, (2) their high modulus makes the reinforced structure stiff, and (3) plastics have good adhesion characteristics towards carbon fibers and can make sound structures. The carbon fibers can be aligned in the required direction or directions and can thus be fabricated as reinforced plastics. There are several methods by which carbon fiber-reinforced polymer (CFRP) structures can be prepared.

CFRP Preparation

The Leaky Mold Technique In this method, weighed quantities of aligned carbon fiber tow are introduced into an excess of a low viscosity and low temperature curing resin contained in a mold. The resin is allowed to wet the carbon fibers. The excess of the resin is then forced out of the leaky mold by applying slight pressure. The gap between the top force and the walls of the mold is kept between two and four thousandths of an inch [15], allowing rapid draining of the resin but preventing extrusion of the carbon fibers. A typical CFRP contains equal parts by weight of the resin and fiber, corresponding to a fiber volume fraction of about 40–43% [15].

Hot Compression Molding This method is generally used for mass production of structures using prepregs. The layers of fibers are impregnated with heat-setting resin in dilute solution and dried to yield prepreg tapes. These prepregs are then cut into the required size and stacked at the required orientation in steel molds and consolidated under heat and pressure. The curing temperature may vary from 90 to 320°C, depending upon the resin matrix [15].

Injection Molding In case of carbon fiber-reinforced thermoplastics the fabrication is usually carried out by injection molding from granules [15].

Filament Winding Technique This technique involves the building up of layers of fibers impregnated with the resin by winding them on a rotating frame. Generally continuous lengths of carbon fibers are used in this technique. The carbon fibers are passed through a bath containing the thermosetting resin and wound around a mandrel in the conventional manner according to a programmed layup pattern, and then cured to yield a plastic structure which is strengthened and stiffened in the desired directions. Filament winding has been used to make aerospace hardware, sporting goods, and components for the automotive industry [16].

Pultrusion Technique The molding and filament winding techniques alone could not meet every design requirement of CFRP, and consequently, an alternative process has been developed [12,33–36] which is capable of preparing elongated structural forms of CFRP with predominantly unidirectional reinforcement. This method involves the formation of impregnated fibers through drawing and consolidation in shaped dies.

The method, known as *pultrusion*, is capable of realizing the high-potential strength and stiffness of the carbon fiber by accurate placement of the reinforcement. The resin-impregnated tows of carbon fiber are drawn through an orifice under tension into a die. Hydrostatic pressure is applied in the orifice constriction to drain the excess resin and to expel the entrapped air. This produces a unidirectionally aligned wet-stage composite with a controlled volume fraction of the fiber and resin, depending upon the cross-sectional areas of the fiber and orifice. The drawing force and drawing rate are determined by the shape, size, volume fraction, and resin viscosity. The wet-stage composite is allowed to stay in the elongated die until gelation, partial curing, and solidification of the composite into the required shape has taken place. Spencer [12] has described pilot-scale pultrusion equipment (Figure 1), which is capable of handling up to 200 tows (1 tow = 10,000 filaments) and giving a range of CFRP profiles. The production rate can range between 70 and 150 mm/min [12].

The versatility of the pultrusion method can possibly be exploited to produce curved and tapered products by incorporating multidirectional reinforcement by integrating tape or filament-winding operations as a continuous process [12]. It is suggested [12] that the pultrusion technique may also be used to process reinforced ceramics and cements or high-temperature polymers such as polyamides, as well as thermoplastics such as polysulfanes.

The pultruded CFRP has been found to have better mechanical properties than the CFRP prepared by the molded technique (Table 4).

CFRP Properties

Carbon Fiber Form Carbon fiber reinforcements can be used as long and continuous tows or in short and fragmented form. They may be dispersed in a thermosetting or thermoplastic matrix with random or directional orientation. However, the properties of the composite will be different depending upon the size of the fibers used and their

Table 4 Comparison of Mechanical Properties Between Pultruded and Molded CFRP

Property	Molded composite (leaky mold technique)	Pultruded composite
Flexural strength (GN m^2)	1.26	1.44
Flexural modulus (GN m^2)	135	148
Interlaminar shear strength (MN m^2)	71.2	83.4
Impact (Izod) (J)	26.6	27.0

Source: Ref. 12.

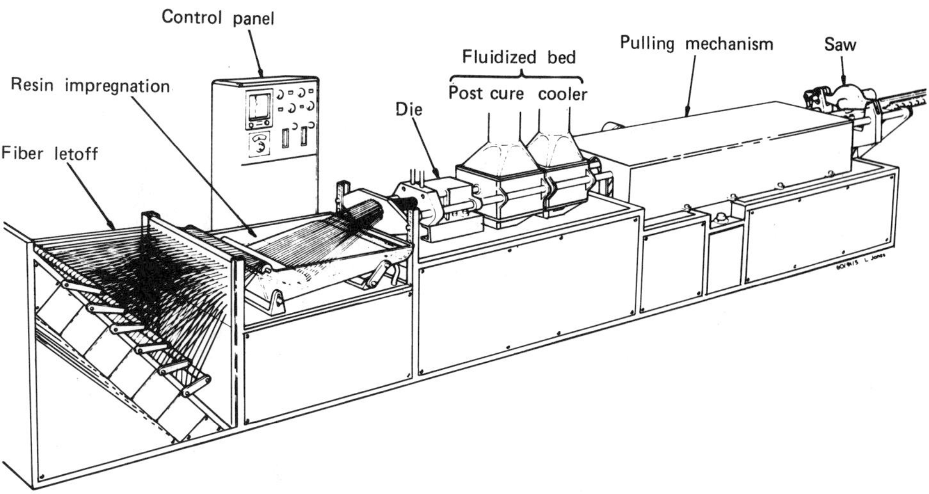

Figure 1 Design of a commercial pultrusion machine. (From Ref. 12.)

Table 5 Effect of Reinforcement Size on the Properties of Reinforced Nylon 6/6 (Injection Molded)

Property	Nylon 6/6	Reinforcement: 30% milled carbon fiber	Reinforcement: 30% chopped carbon fiber
Tensile strength (MN m^2)	77	111	245
Tensile modulus (GN m^2)	2.8	8.9	10.2
Flexural strength (MN m^2)	112	177	357
Flexural modulus (GN m^2)	2.9	10.4	20.2
Compressive strength (MN m^2)	63	98	—
Heat deflection temperature (°C)	70	240	255
Coefficient of thermal expansion (10^{-4} in./in. − °C)	0.81	0.27	0.19
Electrical resistivity (ohm − cm)	10^5	10^3	10^2

Source: Data from Ref. 19.

orientation. Milled fibers, which have an average length of about 300 μm, have also been found to improve the properties considerably [19], although improvement in properties is better when longer fibers are used. Typical properties of reinforced nylon 6/6 prepared by injection molding techniques are given in Table 5. The longer carbon fiber reinforcement significantly increases both the strength and the modulus while maintaining or increasing good electrical conductivity.

Continuous carbon fibers are used as reinforcements when an ultimate in performance and weight reduction is needed and when thermal expansion is to be kept at very low levels. These are available in several forms, such as yarns or tows, containing from 400 to 160,000 individual filaments [19], unidirectional impregnated tapes, plies at selected fiber orientation, and fabrics of many weaves and weights. The choice of the form of the continuous fiber depends upon the fabrication process. For example, filament winding processes generally use lower filament count yarns to minimize looseness of winding, whereas the compression molding processes make use of unidirectional tape.

Table 6 Properties of Different Carbon Fabric-Epoxy Composites

Property	PAN carbon fabric (Thornel 300, Fibrite Corp.)		Pitch carbon fabric	
	Harness satin weave	Plain weave	Harness satin weave (Thornel type P VCB-20)	Plain weave (Thornel type P VCC-45)
Tensile strength (MN m^2)				
Warp	628	559	383	349
Fill	628	559	363	335
Modulus (GN m^2)				
Warp	70	66	42	98
Fill	70	66	41	91
Flexural strength (MN m^2)				
Warp	835	767	489	420
Flexural modulus (GN m^2)				
Warp	66	66	41	93
Shear strength	63	59	31	24

Source: Data from Ref. 19.

Woven Fabric Carbon fibers can also be used in the form of fabrics when complex part shapes and processes require careful orientation of the reinforcement. Carbon fabrics are available in several designs and can have a plain or a satin weave. Commonly used 8-harness satin-weave fabrics retain most of the fiber characteristics in the composite and can be easily draped over complex mold shapes. Plain weave fabrics are less flexible and are suitable for flat or simple contoured parts and have less translation of fiber properties into the composite. PAN carbon fabrics are usually prepared by weaving carbonized yarn, whereas pitch-based carbon fibers are usually woven at an intermediate processing stage and then converted into a high-modulus product while in the fabric form. The properties of carbon fabric—epoxy composites for some fabric materials are given in Table 6. Although the strength properties of pitch-based carbon fabrics are

Table 7 Properties of Plain and Hybridized Composites

Property	All carbon fabric[a]	All glass fabric[b]	Carbon (warp) and glass (fill) fabric[c]
	Reinforcement		
Tensile strength (MN m^2)			
Warp	787	87.7	1325
Fill	390		94
Tensile modulus (GN m^2)			
Warp	88	5.8	124
Fill	52		12.4
Flexural strength (MN m^2)			
Warp	1025	137	1360
Fill	627		200
Flexural modulus (GN m^2)			
Warp	88	5.1	134
Fill	52		12.5

[a]Fibrite Corp. Tech. Bull. Fabric S-type W-184 wim-60% by vol. Thornel 300 in epoxy.
[b]PPG Tech. Service Bull. TS-705-1 70% glass by wt. in polyester.
[c]Fibrite Corp. Tech. Bull. Fabric Stype W-190 with 65% by vol. glass and Thornel 300, in epoxy.
Source: Data from Ref. 19.

not as good as those for polyacrylonitrile (PAN)-based products, the inherent lower cost potential for the pitch materials and their continually improving properties make this product attractive.

Carbon Fiber Hybrids Carbon fibers have low impact resistance and high cost. Therefore, several types of hybrids have been prepared using carbon fibers and the low-cost and high-impact-strength glass fibers. Hybrids are available in several forms. For filament winding or tape layup the secondary reinforcement may be introduced as intermixed fibers within the same layer or as separate layers oriented so as to derive maximum benefits from each material. Commercial fabrics are available which contain a mix of fibers within the warp and/or fill, as well as with carbon fibers in one direction and the secondary fiber in the other. Another rapidly developing hybrid combines continuous reinforcement, either all carbon or mixed fibers, with a chopped secondary fiber, usually glass. This type of hybrid can be molded very much like SMC (sheet molding compound) and the continuous reinforcement need be positioned only in those areas of the part that require stiffening or strengthening. The properties of plain and hybridized composites are compared in Table 7.

Resin Matrix The importance of the matrix in determining the performance properties of high-strength carbon fiber composites is not to be underestimated. It has been examined using three types of epoxy resins: a conventional bisphenol A epichlorohydrin resin (Epikote DX 231) cured with a dicyanodiamine-based curing agent, a precondensate resin (Epikote 210) cured with a boron trifluoride amine complex, and a conventional cycloaliphatic resin (Epikote DX 284) cured with a 4,4 diaminodiphenyl sulfone, by Reader and Morley [36a]. Some of the mechanical properties of these composites, at room temperature, before and after immersion in water for four weeks, are given in Table 8. The composites based on the conventional resin showed good mechanical properties up to 70°C and maintained these properties during prolonged aging at this temperature. The composites prepared with the precondensate resin, however, gave good performance up to 150°C and withstood aging at this temperature. The unconventional resin gave composites with rather higher interlaminer shear strength (ILSS), good retention properties at temperatures between 170 and 180°C, and showed indifference to prolonged aging at 150°C. The composites based on all three resins showed good moisture tolerance.

Several high-temperature plastic matrices have now been developed which improve the temperature range of applications of these composites. Polyimide resins are one such class which can produce carbon fiber-reinforced plastic laminates for use up to 400°C [36b]. These composites have a good thermal resistance which extends the long-term use potential of these composites to higher temperatures (Table 9). These composites can be profitably used as compressor blades and for rotating parts. They are particularly useful in internal engine components because of their good stability in mineral oil.

The properties of unidirectional composites are highly anisotropic [37]. The properties parallel to the fiber are determined to a large extent by the fibers, while the properties perpendicular to the fibers are determined by the matrix and the fiber-resin bond. Therefore, for better transverse properties, laminates are prepared from thin plies of unidirectional material with fibers arranged in different directions in the various plies. Typical mechanical properties for an epoxy resin CFRP unidirectionally reinforced with 60 vol % carbon fibers are given in Table 10. These attractive mechanical properties and the relative freedom to tailor the material to the application offers the engineer a major step forward in strong, stiff, and light materials.

Table 8 Properties of Composites Prepared with Different Resins

Property	Conventional resin (Epikote Dx231)	Precondensate resin (Epikote 210)	Nonconventional resin (Epikote DX284)
Initial Properties at 23°C			
Flexural strength (MN m^2)	1480	1130	1430
Flexural modulus (GN m^2)	131	151	124
ILSS (MN m^2)	72	79	100
Properties after immersion in water for 4 weeks at 23°C			
Flexural strength (MN m^2)	1450	1130	1300
Flexural modulus (GN m^2)	122	153	113
ILSS (MN m^2)	66	74	100

Source: Data from Ref. 36a.

Table 9 Properties of Polyimide Resin/Carbon Fiber Laminates Aged in Air

Aging temperature (°C)	Aging time (hr)	High-modulus fiber[a] Flexural strength ($MN\ m^2$)	High-modulus fiber[a] Flexural modulus ($GN\ m^2$)	High-modulus fiber[a] ILSS ($MN\ m^2$)	High-strength fiber[b] Flexural strength ($MN\ m^2$)	High-strength fiber[b] Flexural modulus ($GN\ m^2$)	High-strength fiber[b] ILSS ($MN\ m^2$)
Room temperature	0	810	153	44	1290	110	90
200°C	25	822	151	49	1340	115	78
	100	816	144	46	1390	112	81
	1000	796	156	34	1570	106	76
300°C	25	782	145	50	1460	113	83
	100	764	148	49	1340	109	77
	1000	617	125	27	—	—	—

[a]Surface-treated 0.52 volume fraction Morganite.
[b]Surface-treated 0.50 volume fraction Courtaulds.
Source: Data from Ref. 36b.

Table 10 Mechanical Properties of CFRP Prepared with Epoxy Resin Containing 60 vol % of Carbon Fiber

Property	Longitudinal	Transverse
Tensile strength (MN m^2)	1700	80
Young's modulus (GN m^2)	130	9
Compressive strength (MN m^2)	1200	270
Strain to failure in tension (%)	1.3	0.9
Strain to failure in compression (%)	0.9	3.0
Shear strength (MN m^2)	110	
Shear modulus (GN m^2)	4.8	

Source: Data from Ref. 37.

5.1.2 Carbon–Carbon Composites (CFRC)

Carbon fibers can also be used to reinforce matrices more brittle than themselves, such as bulk carbon and pyrolyzed phenolic matrices. These matrices have ultimate strains typically 0.5% or smaller. Carbon fibers which break at higher strains do not realize their full tensile strength in such matrices because the matrix fractures before they are fully loaded. However, the fiber reinforcement improves the impact strength and fracture toughness of these composites by arresting cracks which otherwise would propagate through the matrix.

Carbon–carbon composites consist of a fibrous carbon substrate in a carbonaceous matrix. Although both constituents consist of the same element, carbon, the composite behavior is very complex and depends upon the type of reinforcing carbon fiber, the fiber volume fraction, and fiber arrangement on the one hand, and on carbonaceous matrix parameters such as the matrix structure, which may range from carbon to graphite, the processing conditions, and the temperature treatment of the composite, on the other hand.

Preparation of Carbon–Carbon Composites

The continuous carbon filaments are impregnated with an organic matrix precursor, which may be a pitch or a resin [22–26,38–42], by a wet winding technique. The impregnated fibers are heated to remove the solvent and increase the molecular weight of the resin. The fibers are then oriented in the desired arrangement and in the required volume fraction and molded into prepregs which are further pressed and cured into laminates [25]. The laminates are carbonized in an inert atmosphere at 800-1100°C, at rates varying between 5 and 20°C hr^{-1} depending upon the thickness of the material [26]. The weight loss and the volume shrinkage of the organic carbon matrix produces a porous carbon—carbon composite which needs to be densified. The densification of the carbonized composite is carried out either by chemical vapor deposition (CVD) of carbon from a hydrocarbon gas (which is usually methane or benzene), or by multiple impregnations and carbonizations with a liquid organic resin or pitch.

Chemical Vapor Deposition (CVD) The CVD technique is based on the diffusion of a carbonaceous gas into the porous substrate and the diffusion-controlled reaction of the gas with the heated substrate, resulting in the deposition of carbon and elimination of gaseous byproducts. The quality and structure of the resulting deposited carbon depends on the density and surface area of the substrate and on experimental conditions, such as temperature, pressure, and rate of flow of the gas. Several CVD techniques have been developed for the infiltration of the porous substrate, but the isothermal and temperature gradient techniques are more commonly used [25,27—29,40, 42].

In the isothermal technique the substrate is heated by an induction-heated susceptor made of graphite. The hydrocarbon gas decomposes in contact with the hot substrate, depositing carbon. This technique, however, results in overcrusting of the outer surface by deposited carbon, which has to be machined away to achieve maximal density. In general, infiltration requires about 2000 hours [25].

The infiltration by the temperature gradient technique involves shorter times. A nonconducting sleeve, which serves as an insulator and channels the reactants and the byproducts, is placed between a graphite mandrel, which acts as a susceptor, and the induction coil. The porous substrate is not induction-heated owing to its low density and, therefore, has poor coupling. The inner side of the substrate is in direct contact with the mandrel and is, therefore, at a higher temperature. The outer surface of the substrate is in contact with cool gases flowing over its surface. Consequently, the deposition starts at the mandrel and progresses radially through the substrate as the compacted parts of the substrate become inductively heated [25].

Multiple Impregnation Technique The densification of the carbonized composite is carried out in a combined vacuum-pressure impregnation step followed by recarbonization of the impregnated material. The porous composite is first evacuated at 400 Pa and then dipped into the impregnating liquid for about 20 hr under a pressure varying between 1 and 200 Pa, depending upon the precursor material [25]. The recarbonization is then carried out between 1000 and

2880°C. This densification cycle has to be repeated between 4 and 12 times, depending on the thickness and the required final density of the composite [25,30–32].

Properties of Carbon–Carbon Composites

It is essential that the organic precursor which is used as an impregnant should yield a high carbon residue and show a low volume shrinkage in order that the C–C composite have better mechanical properties. Two types of precursor materials have been recognized: thermosetting resins such as phenolics and polyimides, and thermoplastic precursors such as pitches. Phenolic resins and polyimides are easy to handle and give considerable carbon yield (60–65%), but they undergo a high isotropic shrinkage. Consequently, pitches, which under certain conditions [43–46] can yield high carbon contents, are preferred because of their lower shrinkage. Synthetic pitches such as isomeric polyphylenes also give high carbon yields, but they are difficult to handle [47]. The shrinkage of the carbon–carbon composite also depends on the nature of the carbon fiber surface and decreases with decreasing concentration of active surface groups on the carbon fiber surface. The presence of these active surface groups increases adhesion between the fiber and the matrix which, along with high cross-sectional shrinkage, causes pyrolysis cracks in the matrix perpendicular to the fiber axis and leads to shear stress concentration which may result in fiber damage. Fitzer et al. [48] recently studied the effect of carbon fiber surface treatment on the mechanical properties of carbon–carbon composites and observed that a good adhesion, especially the chemical bonds between the matrix precursor and the fiber surface, are detrimental to achieving high flexural strength of the carbon fiber-reinforced carbon (CFRC). The chemical bonding between the liquid matrix precursor and the fiber surface is retained during the carbonization shrinkage of the matrix and causes damage to the fiber-matrix carbon structure. This damage increases with increasing area of chemical bonding. Even densification cycles may not be able to compensate or repair the damage introduced during the first carbonization [48]. It has also been observed [49] that the flexural strength of carbon–carbon composites with phenolic resins can be improved by reducing carbonization shrinkage by the addition of graphite fillers such as pitches of suitable grain.

The mechanical properties of several carbon–carbon composites at room temperature after 4–6 densification cycles and a final heat treatment temperature (HTT) at 1000°C (Table 11), show that the flexural strength and Young's modulus are spread over a wide range. Thus, by selecting the fiber type, fiber volume fraction, fiber arrangement, matrix precursor, processing technique, and the experimental conditions, composites can be tailor made to suit the required applications.

Table 11 Mechanical Properties of Different Carbon-Carbon Composites at Room Temperature

Composite	Fiber content (vol %)	Flexural strength (MN m^2)	Young's modulus (GN m^2)	ILSS (MN m^2)
Unidirectional	55	1200–1400	150–200	20–40
Bidirectional 8H/S cloth wrap	35	300	60	20–40
Three-directional cloth	50	250–300	50–150	50–80
Three-directional felt	35	170	15–20	10–30

Source: Data from Ref. 25.

The thermal expansion behavior of CFRC parallel to the fiber axis is controlled only by the thermal expansion of the carbon fiber, whereas the thermal expansion perpendicular to the fiber axis is determined mainly by the carbonaceous matrix. It is, therefore, possible to make CFRC with excellent high-temperature properties by a suitable choice of carbon fiber and matrix. Consequently, they are mainly used as ablation and high-temperature materials in sophisticated engineering because they have a high heat of ablation, thermal shock resistance, strength improvement, and chemical inertness at high temperatures. Multidirectionally reinforced CFRCs are used in space technology mainly as heat shields and nose tips for reentry vehicles and as rocket nozzles for solid propellants.

5.1.3 Carbon Metal Composites

Most metals are fairly strong, ductile, and isotropic. A considerable amount of effort has been devoted to optimizing these properties by refining, alloying, and heat treatment during the last century and considerable success has been achieved. However, several limitations still exist. One of these is the lack of high-temperature strength which restricts their applications. Consequently, attempts are being made to reinforce metals with carbon fibers to make composites useful for aerospace and general engineering applications. Several metals have been reinforced by high-performance carbon fibers, but the three which have shown the most economic promise are aluminum for aerospace applications, high-temperature oxidation-resistant nickel, and lead-tin alloys as bearing metals. However, there are still problems to be overcome before any of these can reach the stage of commercial exploitation.

Fabrication of Carbon-Metal Composites

The fabrication of metal matrix composites is more difficult than that of the organic matrix systems because higher temperatures and pressures are needed. The use of high temperatures may result in the degradation of the carbon fiber. Several well known methods for preparing metal-metal composites have been used to prepare carbon-metal composites. These methods are broadly classified into (1) diffusion coating [2,50,51], (2) electroplating [7,8,52], and (3) liquid metal infiltration [9,10,53].

Diffusion coating involves flow of metal around the fiber to provide metal fiber contact at temperatures below that of the matrix melting-point under pressure. However, this process is not very suitable for carbon fiber-metal composites because filling of the interstices between very thin carbon fiber filaments in tows is not possible. The pressure involved in diffusion may also cause extensive fiber damage.

Electroplating comprises electrodeposition of matrix material in successive layers to build up the desired thickness of the composite material. However, this process needs subsequent hot processing of the fibers after electrodeposition to obtain dense composites. This may again involve fiber damage.

The liquid metal infiltration technique, which is more commonly used, involves the least fiber damage. The process, however, depends largely on the wetting of the fiber by the liquid metal and on surface tension between the fiber and the liquid to achieve complete penetration of the fiber. However, in certain cases, the liquid metal may react with the carbon fiber, resulting in fiber degradation. In such a case the fibers are first coated with another metal which prevents interaction with the metal matrix and promotes wetting.

Carbon Fiber-Reinforced Aluminum

Carbon—aluminum composites have been prepared by electroplating [52], hot processing of the coated fibers [50], and liquid metal infiltration of bare and coated fibers [9,53]. Composites with good tensile strengths at low and high temperatures have been obtained but very little data are available for more complex mechanical properties such as stress rupture, impact, and fatigue [54]. Paprocki et al. [10] have produced carbon fiber-reinforced aluminum by liquid metal infiltration. Tows of multifilament fibers were coated with a fine layer of titanium and boron by the reduction of titanium tetrachloride and boron trichloride with zinc vapor. This coating activates the fiber surface and promotes wetting and infiltration by molten aluminum, and results in intimate bonding of the fiber to the metal. The composite so ob-

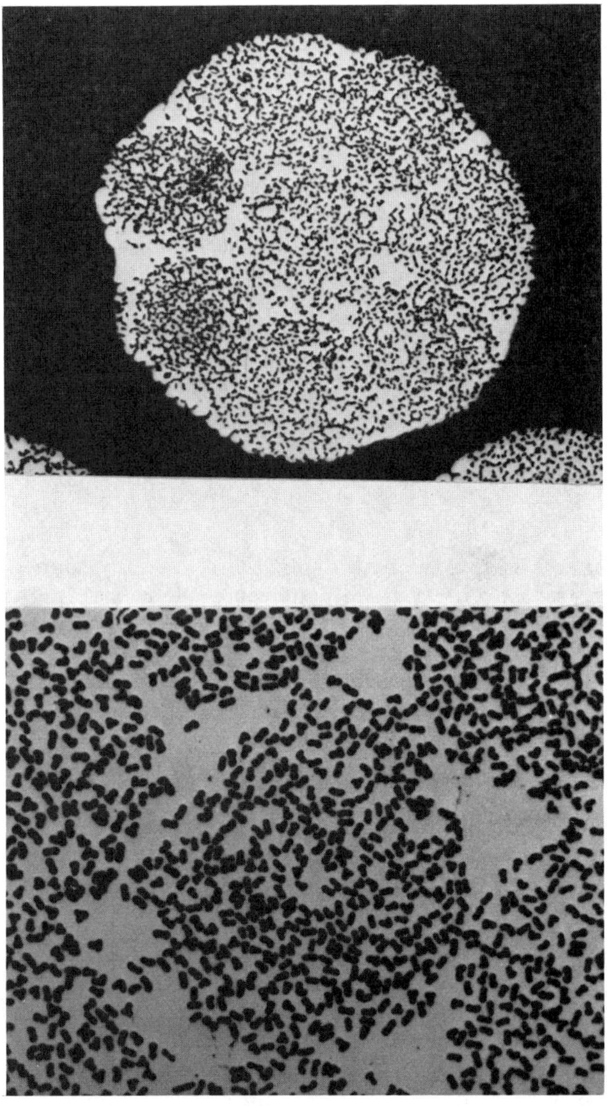

Figure 2 Micrograph of a wire of a carbon fiber-reinforced aluminum composite. (From Ref. 10.)

Table 12 Mechanical Properties of Carbon Fibers and Their Composites with Aluminum

	Carbon fiber		Carbon-aluminum composite	
Property	Celanese GY 70 SE	Union Carbide VS 00 54	Gy 70 SE-201 Al	VS 00 54-201 Al
Tensile strength (MN m^2)	1920	1745–2269	803–837	1045–1080
Tensile modulus (GN m^2)	485–554	693	209	348
Fiber loading vol %	—	—	37–38	48–52

Source: Data from Ref. 10.

tained was found to have a high specific strength and specific modulus in accordance with the rule-of-mixture value determination. Structural studies of the composite material showed a relative uniform infiltration of the metal around the carbon fibers (Figure 2).

The mechanical properties of carbon fibers and their resulting aluminum composites (Table 12) show a good translation of the fiber properties into the composite. The high-modulus carbon fibers did not show any damage due to the formation of carbides and were very stable in aluminum, indicating that these composites should retain their properties at relatively high temperatures. These workers also prepared carbon-magnesium composites which possessed even better mechanical properties [10].

Carbon Fiber-Reinforced Nickel

Carbon–nickel composites have been prepared with good initial mechanical properties [6–8,52a]. These properties, however, deteriorate on exposure of the composites to high temperature or to oxidizing environments. These limitations of carbon–nickel composites have been overcome to a certain extent by a protective coating of the fiber. This, however, involves difficult procedures and is, therefore, not attractive.

Carbon Fiber-Reinforced Lead–Tin Alloy

Reinforced lead–tin alloy metal composites have been obtained generally by infiltration techniques. The infiltration of lead–tin alloy was carried out either by using bundles of copper-plated fibers [55,56], or by immersing fibers at 800°C in tin alloyed with titanium [11]. The composite prepared by taking alloys containing 30–40 wt% lead developed the full strength predicted by the law of mixtures [56]. The tensile strength, however, decreased on either side of this range. On the tin-rich side, the changes in matrix strength with composition and on the lead-rich side the changes in interfacial shear strength cause a decrease in strength of the composite. These composites have good frictional and load-bearing characteristics and can, therefore, be used for bearings.

It is apparent that the carbon–metal composites, when capable of being exploited commercially, will have certain advantages over the epoxy-based composite materials. They have high specific properties, good thermal and electrical conductivity, and exceptional dimensional stability. In addition, they offer the many advantages associated with the metal-based systems. They can be readily clad and joined and attachments can be made with relative ease. The principal difficulty with them at the present is the complexity and the difficulty of preparing them.

5.2 APPLICATIONS

High specific strength and stiffness, together with lightness of carbon fiber composites are the basic characteristics which are of importance in all of their applications. Composites with properties superior to steel and having weights lower by 40–70% have actually been realized. There is a greater degree of flexibility in carbon fiber-reinforced materials in that the strength and stiffness can be varied significantly in different areas of a composite part by selecting the type, form, and suitable orientation of the fiber in the composite, and by controlling the local concentration of the fiber. In addition, several other physical properties of carbon fiber composites such as their thermal stability, electrical conductivity, and corrosion resistance can be varied by a suitable choice of the matrix material and by varying the processing conditions for the preparation of the composite.

We have divided the carbon fiber composite applications on the basis of their most significant physical properties in a particular application, but by no means does this imply that the other physical properties do not play their role.

5.2.1 Applications Based on Strength and Stiffness

Aerospace Applications

The high specific stiffness of CFRP makes them valuable materials for maintaining aerodynamic shapes in wing spoilers and rudders and, combined with high strength, these composites are useful for aircraft wings, tail parts, and helicopter rotor blades [37]. Generally, laminated CFRP structures are used in these aerospace structures. Laminated parts now flying successfully on several aircrafts include landing gear doors of the F-14 fighter (Figure 3), and wing spoilers on more than 100 Boeing 737s in a NASA fleet [16]. The floor decking of the Boeing 747 is also made from laminated carbon fiber prepreg and Dupont's Nemex aramid [16]. The weapon bay doors of the B-1 bomber are also made from CFRP. It is interesting to note that the weapon bay door alone saves about 19% in weight and about 13.5% in cost over the metal door [57]. The high specific strength of carbon-reinforced materials is also made use of in pressure vessels and rocket motor cases.

In addition to the aeroframe itself, weight saving is also important in airborne equipment. The Plassey telescopic cabin (Figure 4A), which is a complete radar installation carried to forward areas by a helicopter, shows a weight savings of about 25% when constructed by using CFRP framework. The CFRP high-speed aerial reflector is another piece of airborne equipment which is 12 times stiffer than the earlier unsatisfactory component, which straightened out under the centrifugal force of the high speed of its rotation [15].

Figure 3 Main landing-gear door made of carbon-epoxy.

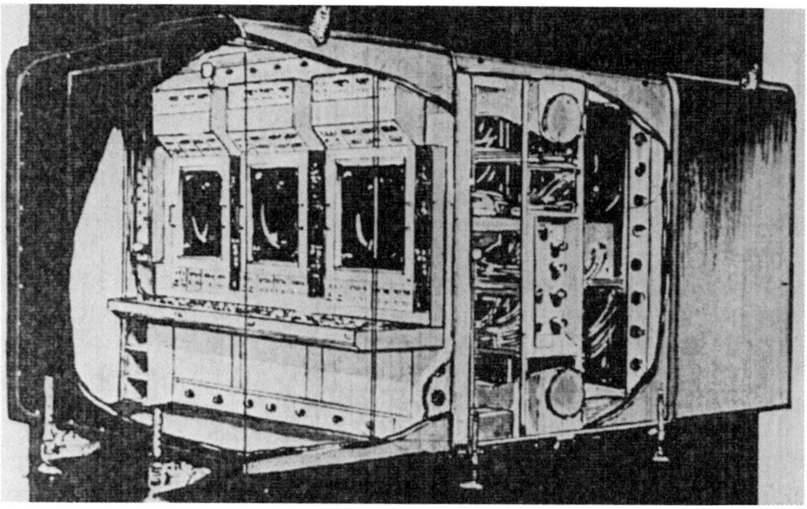

Figure 4A Plassey telescopic cabin. (From Ref. 15.)

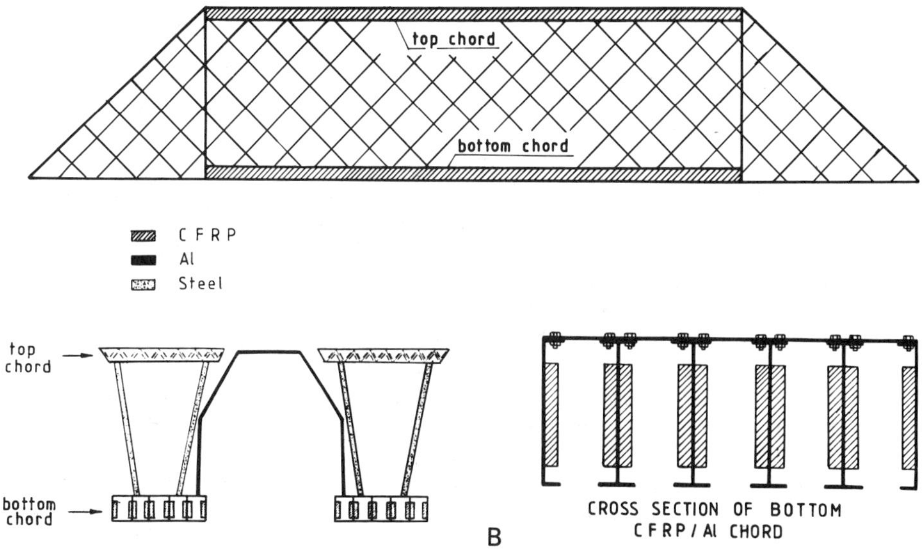

Figure 4B Transportable bridge from CFRP + Al.

Application in Road and Marine Transport

In order to reduce energy consumption, CFRP is being actively considered for reducing the weight of transport vehicles without any loss of strength. Two major load-bearing components which are being particularly examined are the drive shaft [58], and the body spring [59]. The CFRP drive shaft made by wet-filament-binding technique has a weight about half of its steel counterpart with comparable flexural and torsional strength and stiffness. The longitudinal and circumferential stiffness of this shaft can be varied almost independently by changing the yarn winding pattern so that the drive shaft transmits the torque but damps out noise vibrations. The weight of the body spring made from CFRP is almost 25% that of the comparable metal part. The lifetime of these components is also much better because of the high fatigue resistance properties of carbon fibers. The stiffness, rapid recover, and fatigue resistance of CFRP laminates can also be exploited to construct backup beams for bumpers and collision beams in doors, all of which represent weight savings as well [16]. CRFP is also being investigated to affect weight reductions in surface effect vessels and hovercrafts [17], and in making transportable bridges (Figure 4B).

Applications in Sports Goods

The high specific strength and stiffness coupled with rapid damping and good fatigue resistance of carbon fibers make them versatile materials for certain sports articles such as golf club shafts, fishing rods, telescoping casting rods for canal fishing, vaulting poles, ski poles, cricket bats, tennis and badminton racquets, bicycle frames, and racing cars. Filament-wound carbon fiber golf clubs are very popular in the United States, and have consumed the bulk production of the carbon fibers. A golf player is ready to pay the high cost of the club if he can add more distance to his drive at golf. The reinforcement by carbon fibers results in a lighter shaft, permitting proportionally more weight than usual in the head of the golf club. Similarly, carbon fiber-reinforced skis are stiffer and, therefore, require less footpower in cross-country races. Fiber-reinforced laminated bows are stiffer, recover faster, and throw arrows further. Figure 5 shows a Ford GT-40 sports car with a body made from bundles of short-length carbon fibers applied as a grid pattern over the hood, fender, and roof. The addition of carbon fiber reduced the weight of the body (from 153 to 93 lb), increased the stiffness so that the aerodynamic shape was maintained under wind pressure at high speeds, eliminated drumming inside the car caused by roof vibration, and increased the life of the GRP body substantially [60]. This car won the LeMans race several times.

Figure 5 Ford GT 40 sport car. (From Ref. 60.)

5.2.2 Applications Based on Thermophysical Properties

Carbon fibers have exceptional thermophysical properties, which combined with their stiffness, make them extremely useful in composites for higher temperature applications and distinguish them from other reinforcing agents. They have thermal conductivity which helps heat dissipation in components such as gears, bearings, brake disks, and other friction-related products, and in cryogenic processing equipment. They have high thermal stability and are, therfore, the only fibrous material which can be used above 500°C. The low thermal conductivity perpendicular to the fiber axis makes them useful materials for heat shields in missiles [1,18,61].

Heat Shields in Missiles

Heat shields in a missile protect the payload from high temperature, and the severe aerodynamic conditions experienced during the reentry of the missile into the atmosphere. They are generally composed of a relatively brittle phenolic matrix, and a low-modulus carbon fabric [1]. The phenolic matrix pyrolyzes during reentry of the missile, forming a carbon—carbon composite which is eroded by the heat and abrasion. The carbon fiber reinforcement prevents the shield from

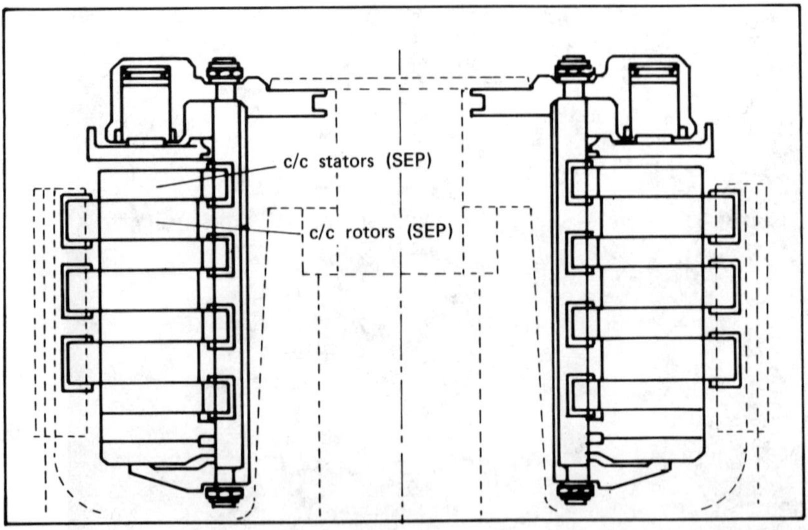

Figure 6 Schematic of an aircraft disk brake. (From SEP Bulletin SEPCARB 1981.)

cracking while its high longitudinal thermal conductivity and low specific heat minimize temperature gradient over the surface, preventing nonuniform erosion of the shield which would have unbalanced the missile. The carbonized heat shield also insulates the payload as the thermal conductivity of the shield is very near normal to the fabric.

Aircraft Brakes

The brakes convert the kinetic energy of a moving vehicle into heat which is stored in the brake material until it can be dissipated. Consequently, aircraft brakes experience very high temperatures. The conventional metal brake materials undergo considerable changes in their strength and frictional properties at these high temperatures and sometimes may even melt. When made from carbon–carbon composites, the aircraft brakes are found to have remarkably stable frictional characteristics to temperatures well above 2000°C. Carbon is stronger at 2000°C than at room temperature, whereas steel melts at 1600°C. The carbon fibers also provide the resistance to thermal shock which is so important in this application. Carbon–carbon composite brakes are lighter because the density of this material is one-fourth the density of steel.

Goodyear has developed CFRC air brakes for F-15 fighter planes which weigh about 30 kg as opposed to 100 kg, and observed that CFRC brakes perform about 3 to 5 times better than their steel coun-

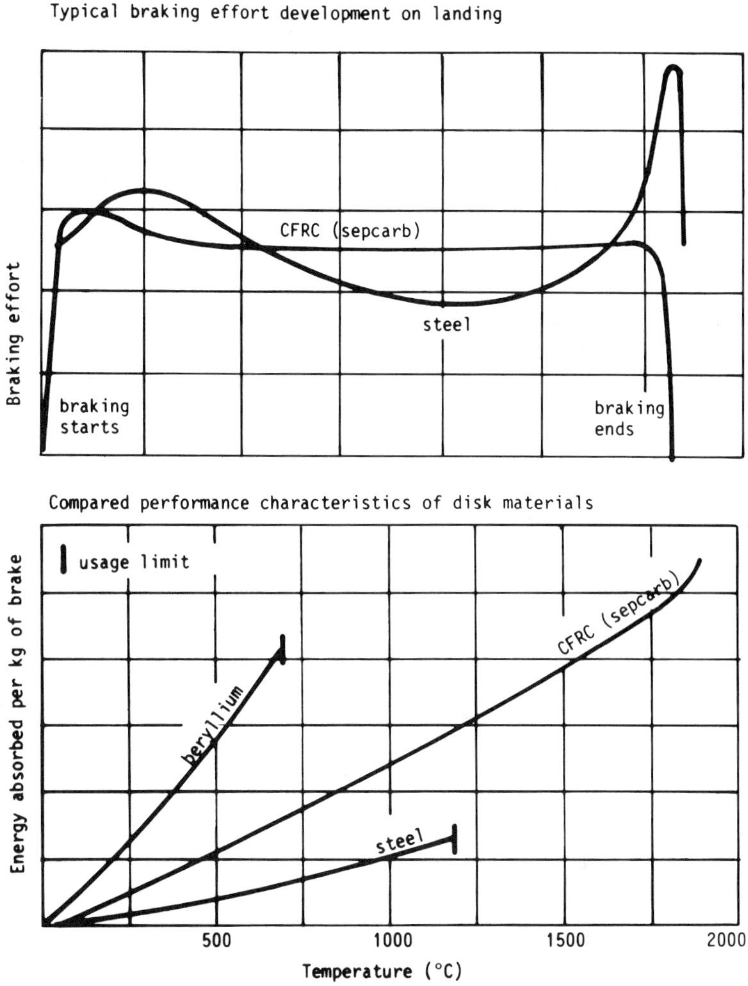

Figure 7 Comparison of performance of a carbon-carbon composite brake and a conventional steel brake. (From SEP Bulletin SEPCARB 1981.)

terparts in actual number of landings before burning out [62]. Thus, the value of C—C composite brakes is not just in weight saving but also in performance. Societé Européenne de Propulsion (SEP) has also developed C—C composite disk brakes for the Mirage 2000 and supersonic Concorde. A schematic diagram of an aircraft disk brake is shown in Figure 6. The typical braking effort on landing and the performance characteristics of brake disk materials are compared in

Figure 7 on an arbitrary scale [63]. The C—C composite brakes are far superior to metal brakes.

Brake disks have been fabricated from both rayon and PAN precursor filaments, using either resin coke or pyrolytic carbon or a combination as the binder material. The PAN filaments give a density increase of about 12% over rayon filaments in similar matrices. This has an added advantage that a smaller physical package can be used for a given kinetic energy absorption [62].

Aerospace Antenna

The low coefficient of linear thermal expansion of carbon fibers makes it possible to design structures with almost zero or very low linear or planar thermal expansion. This is a valuable characteristic for precision instrument components such as telescopes, and for the alignment of aerospace antennas. British Aircraft Corporation (BAC), has developed radar antennas based on CFRP aluminum honeycomb construction capable of retaining dimensional stability in widely different environmental conditions. A 6-ft high carbon sandwich antenna made from an assembly of CFRP laminates bonded to an aluminum honeycomb core built by the Hughes Aircraft Company for the Canadian communication satellite *Anik* is shown in Figure 8. The CFRP components can be used only in a limited temperature range. However, investigations are now underway to develop resin materials which would enhance the temperature range for applications of CFRP. Fitzer and Heym [18] have suggested the use of polyimide matrices for reinforcement by carbon fibers for high-temperature applications. However, these resin polymers are expensive and, therefore, have limited applications. BAC has used carbon fiber-reinforced polyimide type Kerimid 601 for making missile fins [18]. These fins have successfully flown on test missiles. There also appears to be an excellent prospect of using carbon fiber-reinforced polyimide composites in engine service where the temperature may exceed 450°C [16].

5.2.3 Applications Based on Chemical Inertness and Corrosion Resistance

Chemical Industry

Carbon fibers show chemical inertness to a wide range of reagents and hostile environments with exceptional stiffness and lightness. Consequently, carbon fibers in the form of felt, mat, or cloth impregnated by a resin matrix can be used as protective lining for chemical plants and in anticorrosion equipment including bulk liquid transport. A significant application is the construction of acid distribution troughs which are made from carbon fiber felt cladded on GRP and foam polymethane core [59]. The carbon fiber-reinforced troughs are stronger

Figure 8 Communication satellite antenna.

and lighter than the traditional troughs made from asbestos-reinforced phenolic resins for the same cost and do not deform under their own weight. Another important application is the construction of CFRP blades for rotary compressors. This would allow them to run without oil mist lubrication in order to produce oil-free compression [15] (Figure 9). The possibility of fabricating aircraft fuel tanks has already been evaluated [64].

Nuclear Field

Another important area in which high strength and high stiffness coupled with chemical inertness of carbon fibers outweighs the cost disadvantage is in the construction of a gas centrifuge for uranium enrichment. The GRP and metals react with fluorine in uranium hexafluoride involved in the uranium enrichment and, therefore, require a protective coating [17]. Carbon fiber composites need no extra

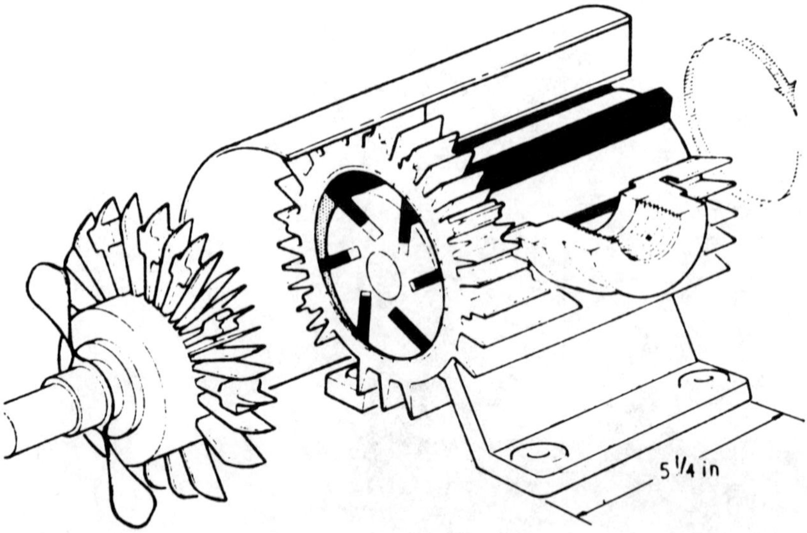

Figure 9 Oil-free compressor with CFRP blades. (From Ref. 15.)

coating and, therefore, result in weight saving without losing strength.

It may, however, be pointed out that although the advantages of improved resistance to acids, alkalis, and reducing agents are real, they cannot be, at present, fully utilized as the resin matrices are usually more vulnerable chemically than the carbon fibers. It is, therefore, necessary to develop greatly improved resins before carbon fiber composites can be used in the chemical industry.

5.2.4 Applications Based on Specific Rigidity and Good Damping

Audio Equipment

Audio equipment requires some parts to vibrate faithfully with sound wave or electrical signals and others to resist vibrations. The vibrating part should be rigid enough to vibrate with the same amplitude and phase and be light enough to respond more effectively to musical and vocal vibrations. In addition, the material should have good damping characteristics to produce pure tones.

Paper or plastic materials which are generally used are light and have good damping, but they are weak. Aluminum, titanium, and other light metals are strong but have inferior damping properties. CFRPs which have all of these properties are, therefore, being extensively used in phonographs and other musical instruments [16,65]. The carbon fiber not only modulates the acoustic properties of plastics, but also dampens the acoustic vibrations most effectively. The audio equipment components which are now being made from CFRP include speaker diaphragms, speaker cone-edges and tone pick-up arms. The light weight of the CFRP pick-up arm makes lateral tracking much more precise and minimizes wear of the record groove. Violins and guitars have also been designed with carbon fiber-filled epoxy laminate sound boards with highly satisfactory results [16].

5.2.5 Applications Based on Electrical Conductivity

Addition of carbon fibers to high-performance plastics makes them electrically conductive. This conductive board can act as a shield against radio frequency interferences. Molded plastic shrouds, made conductive by overlaying a carbon fiber veil mat, can also be used to shield computers and electrical instruments from static or electromagnetic radiations [16]. Carbon fiber-reinforced polycarbonates are used for manufacturing touch switches [59,66]. The ABS switches used inserted metallic conductors. The substitution of polycarbonate has simplified the design and eliminated several secondary operations during the assembly, resulting in appreciable cost savings. A recent innovation is the use of unsupported carbon fibers (not in composite form) in antistatic record brushes, utilizing their inherent electrical conductivity [59].

5.2.6 Applications Based on Biological Inertness and X-Ray Permeability

Medical Applications

The biological inertness, the specific strength, and the pliability of carbon fibers make them useful as replacements for damaged or absent tendons, as surgical implants for regenerating tissues which can grow around, along, and through the fiber bundle [59,67–71] (Figures 10–14). In addition, carbon fibers can be used in x-ray equipment because they are more permeable to x-rays than the conventional metallic components.

Implant materials for endoprostheses were first developed in the 1960s [96], but interest in carbon as an implant material was not raised until 10 years later by the pioneering work of Bokros et al. [97,97a]. Today carbon, in its many forms, is widely used and is considered an interesting implant material for many endoprosthetic devices. Hüttinger has recently given a very good review of its uses [98,98a]; a complete review of biomedical applications of carbon and graphite through 1980 has been summarized by Jenkins [99]. For example, in recent years more than 400,000 carbon components for heart valves have been implanted.

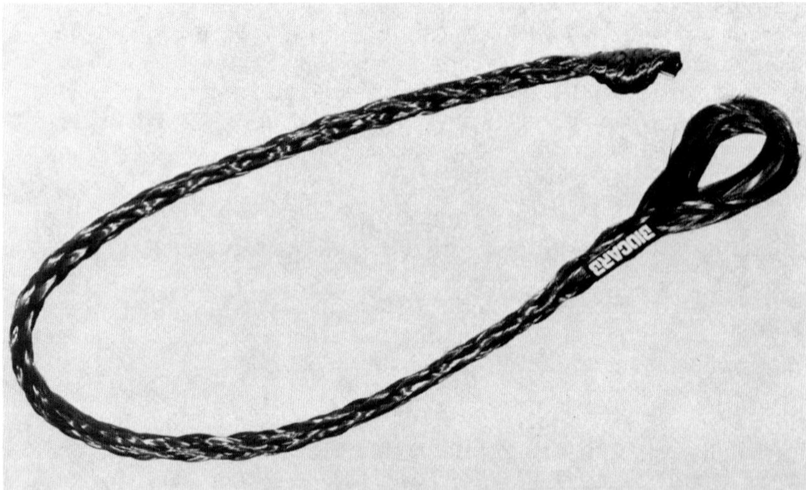

Figure 10A Carbon fiber prosthetic for tendo Achilles. (From SEP Bulletin SEPCARB 1981.)

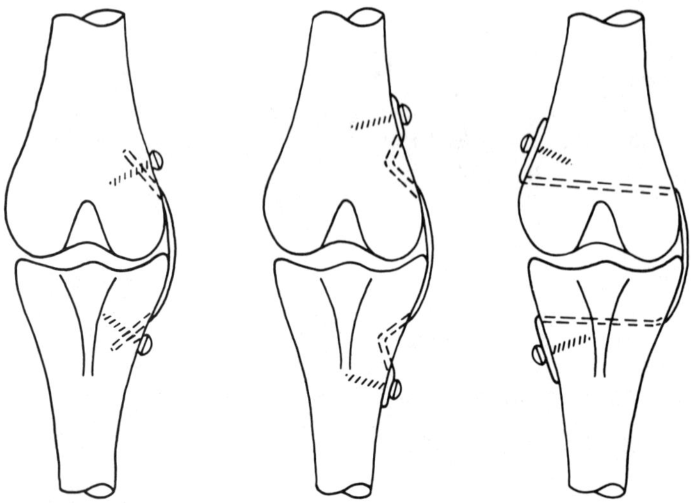

Figure 10B Carbon fiber prosthetics for joints. (From SEP Bulletin SEPCARB 1981.)

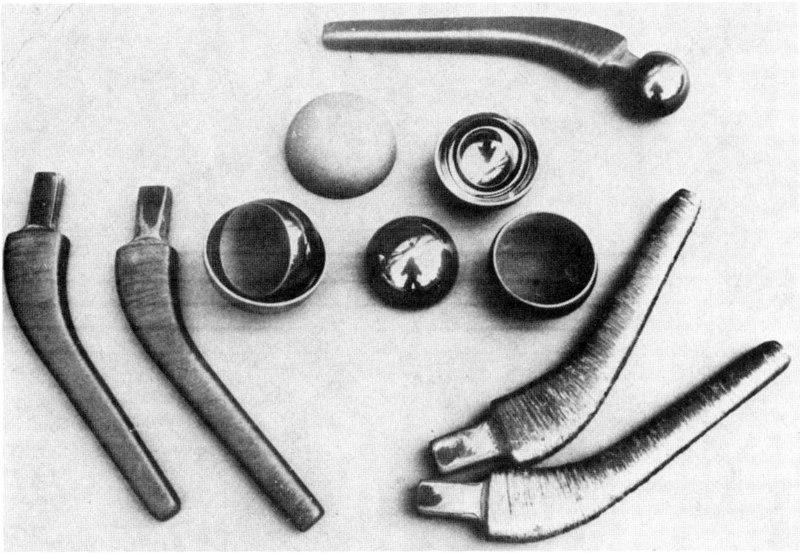

Figure 10C Carbon fiber prosthetics for joints. (From SEP Bulletin SEPCARB 1981.)

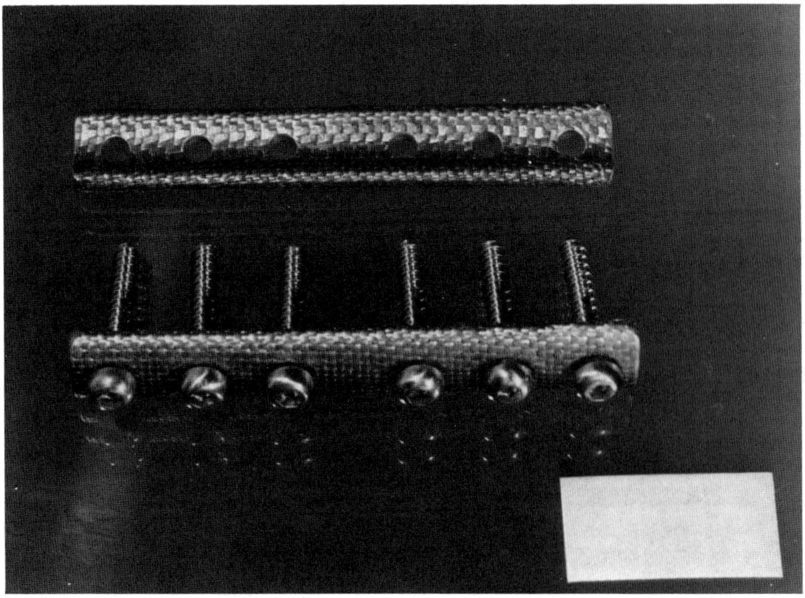

Figure 11 Carbon fiber composite for bone joints. (From SEP Bulletin SEPCARB 1981.)

Figure 12 Carbon fiber composite hydraulic motor for artificial heart implant. (From SEP Bulletin SEPCARB 1981.)

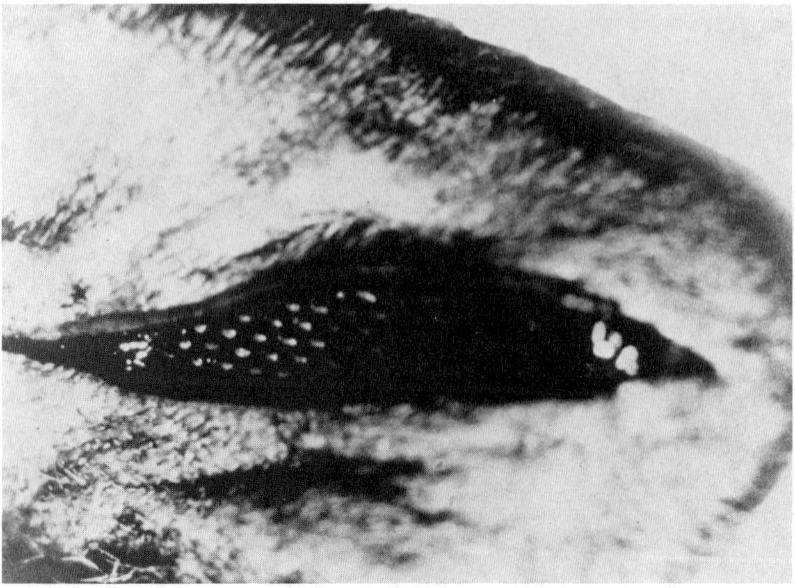

Figure 13 Carbon fiber prosthetic for dog's ear. (From SEP Bulletin SEPCARB 1981.)

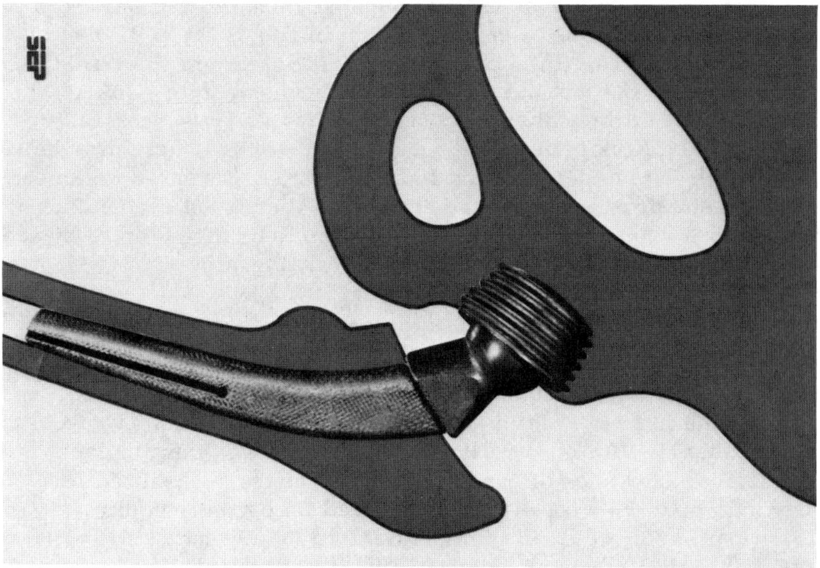

Figure 14 Hip joint prosthesis. (From SEP Bulletin SEPCARB 1981.)

Surgical Implant The first successful attempts to replace tendon and ligament were reported by Jenkins [67,68]. In a series of experiments the tendo achilles of a number of sheep were excised and replaced with a double-plaited strand of flexible carbon fiber tow (Figure 10A). In a controlled series of sheep the tendo achilles was also excised and either replaced with nylon or not replaced at all. The control sheep never walked satisfactorily while the sheep with the inserted carbon fiber walked normally within 2 to 3 weeks. The histological examination of sections following implantations showed no more than the inflammatory reaction expected in a wound. The strength of the implant increased with time and corresponded with the stimulation of an ingrowth of new tissues into the carbon fiber scaffold [68]. The carbon fiber was also found to disappear from the site of the implantation.

Fitzer et al. [71] attribute this disappearance of implanted carbon fiber to mechanical degradation before the tissue is regenerated. They coated carbon fibers with pyro-carbon to make them more resistant to splitting and breaking. The improved carbon fibers were implanted in 20 rabbits. Histological examination after three months, however, showed that fragmentation of the carbon fiber could not be completely prevented.

Although implantation of carbon fibers is still in the developmental stage, it appears that the inertness, the strength of carbon fibers, and the fact that new cells can grow against them in damaged areas during healing will make them very useful in this area. Fitzer et al. [74] have also indicated the possibility of using carbon fiber-reinforced composites as plates for internal fixation of bones (Figure 10B,C).

Prosthetics Carbon fiber-glass hybrids have been successfully used for making prosthetics (Figures 11-14) [59,69]. These prosthetics, although more expensive, are light and strong compared to the conventional metal prosthetics. Furthermore, in the case of carbon fiber-reinforced prosthetics, the pattern of the limb to be fitted can be directly molded which considerably facilitates the fitting process.

X-Ray Equipment

CFRP x-ray pressure plates have an x-ray absorbancy of about one-sixth that of similar aluminum plates and can, therefore, give better resolution in x-ray pictures [59]. The high permeability to x-rays, the high stiffness to resist deflection, and high oscillation damping characteristics of carbon fiber composites are utilized for constructing patient's coach in computerized tomography equipment and for cantilever tables in therapeutic x-ray equipment [59]. Filament-wound CFRP rings are also used in the tomography scanner as a cover for the x-ray tubes. The possibility of using carbon fiber composites as collimeters and filters for elastic waves, useful to construct ultrasonic focusing devices for medical and surgical applications, has also been indicated [17].

5.2.7 Applications Based on Fatigue Resistance, Self-Lubrication, and High Damping

Applications in Textile Machinery

Textile machines involve a combination of high speed reciprocation, oscillation, and rotatory motion where inertia, strength, fatigue, and natural frequency of metal components limit machine operating speeds [72,73]. Since CFRPs provide advantages in these properties, they are extensively being used to make textile machine components. The common components which are being made from CFRP are heeled frames, picking sticks, slays in weaving looms, knitting bars, and push rods in knitting machines. These components are generally made from protruded carbon fiber—epoxy strip and from injection-molded carbon fiber-reinforced nylon [59]. The high thermal stability and the high specific stiffness requirement of a looper bar in a lockstitch machine can only be met by CFRP.

CFRP has self-lubricating properties. Against certain counterface materials and under certain environmental conditions they can transfer a graphite film to preserve both the CFRP and the counterface material against further wear. One application of this property invol-

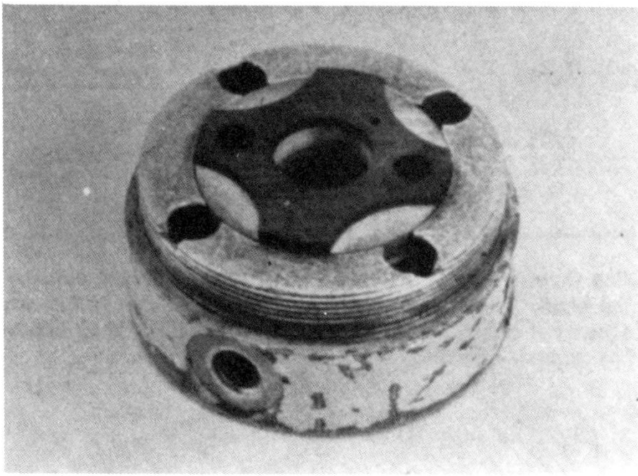

Figure 15 Air slide valve for a loom made with CFRP. (From Ref. 73.)

ving carbon fiber-epoxy composite is in an air slide valve for a loom [73] (Figure 15). The original material used was cast iron against a seating of hardened steel or mirror finish, which needed external lubrication. Substitution with bronze gave some improvement, but external lubrication was still needed. An insert of GRP faced with a thin layer of CFRP has been successfully used for 2 years without any external lubrication and with very little wear [73].

Applications in General Engineering

Several general engineering components such as bearings, gears, cams, and fan blades, and certain components for domestic electrical equipment are also made from carbon fiber composites. These components are generally made from injection-molded carbon fiber-reinforced thermoplastics such as nylon. Compressor blades for oil-free compressors are made using thermosetting matrices [59].

5.3 ACTIVATED CARBON FIBERS

The production of highly effective fibrous carbon adsorbents with low diameter, excluding or minimizing external and intradiffusional resistance to mass transfer, and, therefore, exhibiting high sorption rates, is a challenging task for investigators in the field of science and technology of carbon fibers. These carbon adsorbents can be converted into a wide variety of textile forms and nonwoven materials, with low hydrodynamic resistance, to be used in thin layers for treatment of high gas flows. These materials will increase efficiency and permit a far greater flexibility and simplification in the design of sorption processes for environmental pollution control. The use of carbon fiber adsorbents can also help eliminate problems caused by chanelling and settling of activated carbon powders. Thus, efforts have been under way for almost a decade to prepare cheaper activated carbon fibers with a uniform pore size distribution and high surface area.

Economy and Lin [75,76] prepared activated carbon fibers in 60% yield from a phenolic precursor fiber Kynol in a two-stage process, which involved carbonization and activation. The carbonization was carried out in nitrogen atmosphere to about 800°C at a rate of 100°C/hr. The activation was carried out in steam between 750 and 1000°C. The degree of activation and, consequently, the surface area of the final product depended upon the temperature and the time of activation. Activated carbon fibers with surface area as high as 3000 m^2 g could be obtained.

The adsorption data [77] showed that the activated carbon fibers had pore radii between 5 and 14 Å and displayed excellent adsorption capabilities for liquid and gas pollutants such as phenol, butane, HCN, H_2S, and NO_2. Isotherm studies showed that activated carbon fibers were more effective compared to granulated carbons in removing trace contaminants such as phenol and pesticides from aqueous solutions. This improved efficiency was attributed to a more uniform pore size distribution and a more efficient contact surface.

The bulk of the research in different countries, however, has been aimed at obtaining activated carbon fibers from cellulosic raw materials, which are the cheapest for this class of carbon fibers. Thus, activated carbon fibers have been obtained by the carbonization of cellulosic fibers treated with zinc, aluminum, magnesium, or iron halides followed by activation in CO_2 or in steam in the temperature range 600–1000°C [78–80]. In the U.S.S.R., the activated carbon fibers have generally been obtained by carbonizing tows, fabric, or nonwoven, exregenerated cellulose between 600 and 1200°C and then activating in CO_2 or steam between 600 and 1100°C to a degree of activation varying between 12 and 72% [81-88].

Activated carbon fibers in diverse textile forms are increasingly used in the form of various air filters [88,89,90], gas masks [89], air conditioners [88], and in appliances for water treatment [88,89]. The capacity of a cigarette filter from activated carbon fiber [93] is reported to be more than double that of the filters made from cellulose acetate. Activated carbon fibers are also used for the production of protective clothing used in various chemical industries for work in extremely hostile environments [88,90,92]. They have been reported to be very promising as catalyst carriers [89–94] for platinum, palladium, ruttenium, and rhodium in metallic form or as compounds. Activated carbon fiber ion exchanges have been developed at the Institute of General and Inorganic Chemistry of the Belorussian Academy of Sciences [95].

5.4 CONCLUSION

Carbon fibers, although a relatively new material, have attracted worldwide attention and hold great promise as a future material. Their major use is anticipated in the form of reinforced structural materials such as CFRP, CFRC, and CFRM. "Lighter than aluminum and stronger than steel" has become a cliche for carbon fiber-reinforced composite materials; it is now being realized in practice.

The manufacture of and research into carbon fibers has been directed at several types of starting materials, including rayon. However, it has been narrowed down to two main types: PAN-based and pitch-based carbon fibers. Both fibers were pioneered in Japan, the PAN-based by Shindo and the pitch based by Otani. At present PAN-based carbon fibers which prove highly desirable in high-performance composites for aircraft, aerospace, and similar applications account for the bulk of the production. However, the unique structure, characteristics, and, in some cases the lower price of mesophase pitch-based fibers have opened up a variety of new applications. Consequently, greater emphasis is now being placed on the development of new pitch materials yielding better carbon fibers, and on the development of processes to obtain better materials from existing sources.

Several aspects of carbon fiber structure have been examined during the last two decades and considerable agreement has been reached concerning the various microstructural features. There is general agreement that carbon fibers are composed of two-dimensional graphitic layers, oriented preferentially parallel to the fiber axis. It is also agreed that carbon fibers have a two-phase structure and several kinds of micro- as well as macrostructural heterogeneities. The observation that carbon fibers prepared from different precursor materials have many common structural features is another remarkable feature of these structural studies. However, better and more sophisticated techniques will continue to be used.

Carbon fibers with excellent mechanical properties can now be prepared, but no efficient translation of these outstanding mechanical properties into usable structural form (such as composites) has yet been achieved. Carbon fibers show insufficient adhesion and weak bonding towards most thermosetting resin matrices, resulting in poor interlaminar shear strength of the resulting composite. Several types of surface treatments which include oxidation, coating, and grafting of the carbon fiber surface have been investigated for improving adhesion and interfacial bonding between the carbon fiber and the resin matrix. The oxidative treatments in aqueous solutions which are mild and do not cause excessive burnoff and degradation of the carbon fiber surface have been found to be more suitable. From the point of view of their applicability in the industrial production of carbon fibers, electrochemical oxidations (in aqueous alkaline or acidic solutions) appear to be more practicable. Oxidative treatments increase the specific surface, thus providing more bending sites, and they create surface functionality which enhances the wettability of the fiber by the matrix material. Thus there is need for continued research in the development of these surface treatments and for measurements of the magnitude and nature of the chemical surface groups developed by these treatments.

Noteworthy characteristics of carbon fibers which make them versatile materials include low density, outstanding heat resistance, particularly in vacuum or inert atmosphere and dimensional stability, high electrical and thermal conductivity, low thermal expansion coefficient, and resistance to corrosion from virtually all acids, alkalis, and organic solvents, excluding strong oxidizing agents. The application of carbon fibers which 10 years ago were limited and almost entirely restricted to the aerospace industry, are now spreading to every human endeavor, from toys to human body prosthetics.

The future of carbon fibers is under constant evaluation, not only by producers, sellers, and users, but also economists, for their strategic importance. The published data have been reappraised many times, oscillating from overenthusiasm to gloom. At present there is a seeming worldwide frenzy of announcements concerning new openings. It remains to be seen if all these projects will ever be achieved. The demand for carbon fibers has displayed a remarkable growth of about 50% annually in recent years. The forecasts for 1983 on market development concerning carbon fibers differ widely from 4000 to 22,000 tons per year (in 1990), a difference of a factor of 5.5.

In our opinion, carbon fiber performance and prices will continue to undergo changes, and accordingly, markets will continue to grow. Apart from their rising economic importance, carbon fibers are here to stay and will certainly remain for a long time a fascinating field for scientists and technologists.

REFERENCES

1. Towne, M. K., and Dowell, M. B. Papers of the American Association for Textile Technology, Inc., Modern Textile, May 1976, p. 51.
2. Sambell, R. A. J., Phillips, D. C., and Bowen, D. H. *Carbon Fibers—Their Place in Modern Technology, Proc. 2nd Intern. Carbon Fibre Conf.*, The Plastics Institute, London, 1974, p. 105.
3. Sambell, R. A. J., Bowen, D. H., and Phillips, D. C. J. Mater. Sci. 7, 663 (1972).
4. Phillips, D. C., Sambell, R. A. J., and Bowen, D. H. J. Mater. Sci. 7, 1454 (1972).

5. Phillips, D. C. J. Mater. Sci. 7, 1175 (1972).
6. Barclay, R. B., and Bonfield, W. J. Mater. Sci. *6*, 1076 (1971).
7. Jackson, P. W., and Marjoram, J. R. J. Mater. Sci. 5, 9 (1970).
8. Galano, F. S., and Pinto, J. Fibre Sci. Tech. *2*, 303 (1970).
9. Morris, A. W. H. Carbon fibres—Their composites and applications, *Proc. 1st Intern. Carbon Fibre Conf.*, The Plastics Institute, London, 1971, p. 119.
10. Paprocki, S., Kizer, D., and Meyerer, W. Enigma Eighties Econ. Energy Book 2, 24(2), 1451 (1979).
11. Rashid, M. S., and Wirkus, C. D. Bull. Amer. Cer. Soc. *51*, 836 (1973).
12. Spencer, R. A. P. Carbon fibres—Their place in modern technology, *Proc. 2nd Intern. Carbon Fibre Conf.*, The Plastic Institute, London, 1974, p. 140.
13. Wrzesien, A. P. Carbon fibres—Their place in modern technology, *Proc. 2nd Intern. Carbon Fibre Conf.*, The Plastics Institute, London, 1974, p. 161.
14. Morgan, P. E. Carbon fibres— Their place in modern technology, *Proc. 2nd Intern. Carbon Fibre Conf.*, The Plastics Institute, London, 1974, p. 154.
15. Phillips, L. N. Proc. Instn. Mech. Engrs. 185 52/71, 793 (1970-71).
16. Dowell, M. B. Plastics Engineering p. 31, April (1977).
17. Tewary, V. K. Proc. Nucl. Phys. Solid State Phys. Symp. *18A*, 15 (1975).
18. Fitzer, E., and Heym, M. Chemistry and Industry London p. 663 (1976).
19. Shepler, R. E., Towne, M. K., and Saylor, D. K. Machine Design, p. 3, May (1979).
20. Jeffries, R. Nature *232*, 304 (1971).
21. Delhaes, P., and Trinquecoste *15th Bienn. Conf. on Carbon*, Philadelphia, Penn., 1981, extended abstracts, p. 268.
22. Fitzer, E., Hüttner, W., and Manocha, L. *14th Bienn. Conf. on Carbon*, Pennsylvania State University, University Park, Penn., 1979, extended abstracts, p. 240.
23. Brussell, G. W., and Wei, G. C. *14th Bienn. Conf. on Carbon*, Pennsylvania State University, University Park, Penn., 1979, extended abstracts, p. 247.
24. Markovic, V., Marinkovic, S., and Marsh, H. *15th Bienn. Conf. on Carbon*, Philadelphia, Penn., 1981, extended abstracts, p. 272.
25. Fritz, W., Hüttner, W., and Hartwig, G. *Proc. ICMC Symposium 1978*, Met. Mater. Compos. Low Temp., 1979, p. 245.

26. Hill, J., Thomas, C. R., and Walker, E. J. Carbon fibres—Their place in modern technology, *Proc. 2nd Intern. Carbon Fibre Conf.*, The Plastics Institute, London, 1974, p. 122.
27. Diefendorf, R. J. *Proc. Carbon Comp. Techn. Symp.*, 10th Ann. ASME Symposium, New Mexico, 1971.
28. Kotlensky, W. V. *Proc. 16th Nat. Sampe Symposium and Exhibition*, Anaheim, California, 1971.
29. Stoller, H. M., and Frye, E. R. Sampe Quarterly 3(3), 10 (1971).
30. Fitzer, E., and Terwiesch, B. Carbon *11*, 570 (1973).
31. Evangelidis, J. S. *13th Bienn. Conf. on Carbon*, Irvine, Calif., 1977, extended abstracts, p. 16.
32. Weber, G. W., Young, K. R., and Taylor, A. J. *13th Bienn. Conf. on Carbon*, Irvine, Calif., 1977, extended abstracts, p. 68.
33. Paton, W., Lockhart, A. H., and Montgomery, I. W. Carbon fibres—Their place in modern technology, *Proc. 2nd Intern. Carbon Fibres Conf.*, The Plastics Institute, London, 1974, p. 148.
34. Patton, W. Composite 3(3), 119 (1972).
35. Hancox, N. L., Garnet, G., and Spencer, R. A. P. 3rd Plastic Institute Conf. on Research Projects London (1971).
36. Hardesty, E. E. SPI 28th Reinforced Plastics Tech. Conf. (1973).
36a. Reader, C. E. L., and Morley, J. M. *Proc. Fourth London Intern. Carbon and Graphite Conf.*, 1976, Soc. Chem. Ind. 1976, p. 238.
36b. Rogers, R. F., and Kingston-Lee, D. M. *Proc. 3rd Conf. on Ind. Carbon and Graphite*, Soc. Chem. Ind., 1971, London, p. 503.
37. Dorey, G. Phys. Technol. *11*, 56 (1980).
38. Bradshaw, W. G., and Pinoli, P. C. *12th Bienn. Conf. on Carbon*, Pittsburgh, Penn., 1975, extended abstracts, p. 297.
39. Brunnel, L. R. *12th Bienn. Conf. on Carbon*, Pittsburgh, Penn., 1975, extended abstracts, p. 333.
40. Pacault, A., Loll, P., Delhaes, P., and Pierre, A. *13th Bienn. Conf. on Carbon*, Irvine, Calif., 1977, extended abstracts, p. 164.
41. Dacie, B., and Marinkovic, S. *Carbon '80 3rd Intern. Conf.*, Baden-Baden, 1980, preprints, p. 613.
42. Kimura, S., Takase, N., Kasuya, S., and Yasuda, E. *Carbon '80 3rd Intern. Carbon Conf.*, Baden-Baden, 1980, preprints, p. 617.
43. Fitzer, E., and Bürger, A. Chem. Ing. Tech. *42*, 1203 (1970).

44. Fitzer, E., and Terweisch, B. Carbon *10*, 383 (1972).
45. Fitzer, E., Heym, M., and Rhee, B. *Carbon '76 2nd Intern. Carbon Conf.*, Baden-Baden, 1976, preprints, p. 508.
46. Fitzer, E., and Tillmanns, H. *12th Bienn. Conf. on Carbon*, Pittsburgh, Penn., 1975, extended abstracts, p. 217.
47. Fitzer, E., and Greiser, F. *Carbon '76 2nd Intern. Conf.*, Baden-Baden, 1976, preprints, p. 513.
48. Fitzer, E., Geigl, K. H., and Hüttner, W. *Proc. 5th London Intern. Conf. on Carbon and Graphite*, 1978, Soc. Chem. Ind., London, 1978, p. 493.
49. Fitzer, E., Geigl, K. H., and Hüttner, W. *14th Bienn. Conf. on Carbon*, Pennsylvania State University, University Park, Penn., 1979, extended abstracts, p. 236.
50. Lalacona, F. P. N71-34502 (1971) (Quoted in C. F. Old and M. G. Nicolas paper in Carbon Fibres— Their Place in Modern Technology, the Plastics Institute, London, 1971, p. 89).
51. Denham, A. E., and Redfern, B. A. W. Ger. Patent 2,200,239.
52. Baker, A. A. et al. Composite *2*, 154 (1971).
52a. Wang, W. L., Zhang, G. D., Jin, C., and Wu, R. J. *Intern. Symp. on Carbon*, Toyohashi, Japan, 1982, p. 447.
53. Pepper, R. T. et al. Met. Trans. *2*, 117 (1971).
54. Old, C. F., and Nicholas, M. G. Carbon fibres— Their place in modern technology, *Proc. 2nd Inter. Carbon Fibre Conf.*, The Plastics Institute, London, 1974, p. 89.
55. Howlett, B. A., Minty, D. C., and Old, C. F. Carbon fibres— Their composites and applications, *Proc. Intern. Carbon Fibre Conf.* 1971, The Plastics Institute, London, 1971, p. 99.
56. Old, C. F., and Nicholas, M. G. Carbon fibres— Their place in modern technology, *Proc. 2nd Intern. Carbon Fibre Conf.*, The Plastics Institute, London, 1974, p. 93.
57. Forsyth, R. B., 20th National SAMPE Symposium and Exhibition, San Diego, Calif., April 29-May 1, 1975.
58. Thompson, A. W. New fibres and their composites, Phil. Trans. Roy. Soc. London, *A294*, 577 (1980).
59. Green, A. K., and Phillips, L. N. Materials in Engineering Applications *1*, 59 (1978).
60. Lovell, D. R. Carbon fibres— Their place in modern technology, *Proc. 2nd Intern. Carbon Fibre Conf.*, The Plastics Institute, London, 1974, p. 283.
61. Laramee, R., Lamere, G., Prescott, B., Mitchell, R., Scottosanti, P., and Dahle, D. *13th Bienn. Conf. on Carbon*, Irvine, Calif., 1977, extended abstracts, p. 74.

62. Kirkhard, F. P. *12th Bienn. Conf. on Carbon*, Pittsburgh, Penn., 1975, extended abstracts, p. 279.
63. Sepcarb Carbon/Carbon Composites Materials Activity Bulletin, 1981 Société Européenne de Propulsion Puteaux, France.
64. Van Auken, R. L. 27th SPI Reinforced Plastics/Composites Conf. Washington, D.C. (1972), 13-D.
65. Murayama, K. Jee p. 38 (July 1976).
66. Naitove, M. H. FRP '78: All Eyes on Automotive Plastics Technology, *24*, 67 (1978).
67. Jenkins, D. H. R. Intern. Biomaterials Symposium, Philadelphia, Penn., 1976.
68. Jenkins, D. H. R. *13th Bienn. Conf. on Carbon*, Irvine, Calif., 1977, extended abstracts, p. 178.
69. Johnson, G. R. Conf. on Designing with Fibre Reinforced Materials Instt. of Mech. Engrs., London, 1977, paper C2333/77.
70. McKenna, G. B., and Statton, W. O. J. Appl. Polymer Sci.: Appl. Polymer Symp. *31*, 335 (1977).
71. Fitzer, E., Hüttner, W., and Wolter, D. *13th Bienn. Conf. on Carbon*, Irvine, Calif., 1977, extended abstracts, p. 180.
72. Trewin, E. M. Reinforced Plastics Congress, British Plastics Federation, Brighton, 1976, paper 23.
73. Lucas, R. Carbon fibres—Their place in modern technology, *Proc. 2nd Intern. Carbon Fibre Conf.*, The Plastics Institute, London, 1974, p. 266.
74. Fitzer, E., Hüttner, W., Claes, L., and Kinzl, L. Carbon *18*, 383 (1980).
75. Economy, J., and Lin, R. Y. J. Mater. Sci. *6*, 1151 (1971).
76. Lin, R. Y., and Economy, J. *Applied Polymer Symposia No. 21*, John Wiley, New York, 1973, p. 143.
77. Economy, J., and Lin, R. Y. *Applied Polymer Symposia No. 29*, John Wiley, New York, 1976, p. 199.
78. Br. Patents 1,301,101 (1972), 1,310,011 (1973).
79. U. S. Patents 3,849,332 (1972), 3,847,833 (1974).
80.F. Bolwell, A. S. Design News *28*, 65 (1973).
81. Fridman, L. I. et al. Izv. Akad. Nauk. Belorus S. S. R. Ser. Khim Nauk *no. 2*, 37 (1974).
82. Ermolenko, I. N., Morozowa, A. A., and Gavrilov, M. Z. Doklady Akad. Nauk Belorus S. S. R. *T. 18 no. 3*, 234 (1974).
83. Fridman, L. I., Morozova, A. A., and Ermolenko, N. I. Doklady Akad. Nauk Belorus S. S. R. *T. 18 no. 12*, 1087 (1974).
84. Ermolenko I. N. et al. Izv. Akad. Nauk. Belorus. S. S. R. Ser. Khim. *no. 5*, 20 (1975).

85. Fialkov, A. S. et al. Vysokomol. Soedim Ser. *BT*. *9 no. 6*, 464 (1967).
86. Krasnov, E. P. et al. Vysokomol. Soedin. Ser. A *T*. *12 no. 8*, 1706 (1970).
87. Fridman, L. I., Morozova, A. A., Perlin, V. A., Levit, R. M., Ermolenko, I. R., Konkin, A. A., and Maslak, N. V. Khim Volokna *no. 1*, 11 (1977).
88. Arons, J. N., and MacNair, R. N. Text. Res. J. *44*, 874 (1974).
89. Swiss Patent 514,500 (1970).
90. Arons, J. N., MacNair, R. N., and Erikson, R. L. Text. Res. J. *43*, 539 (1973).
91. Weber, S. Textilveredlung *89*, 412 (1971).
92. U. S. Patent 3,744,534 (1973).
93. Br. Patent 1256048 (1970).
94. Fr. Patent 2,198,893 (1974).
95. Ermolenko, I. N., Sviridova, R. N., and Kuvaeva, Z. I. U. S. Patent 3,702,832 (1972).
96. Charuley, J. Triangle 4, 175 (1960).
97. Bokros, J. C. et al., Chem. Phys. Carbon, *9*, 103 (1973).
97a. Bokros, J. C. Carbon, *15*, 355 (1977).
98. Hüttinger, K. J., Extended Abstracts of Int. Symp. on carbon, Toyohashi, Japan, 1982, p. 29.
98a. Hüttinger and al., Biomaterials *1*, 67 and 73 (1980) and 6th London Int. Carbon and Graphite Conf. (1982).
99. Jenkins, G. M. Chemical Phys. Physiol. Meas., *1*, 171 (1980).

Author Index

Numbers in brackets are reference numbers and indicate that an author's work is referred to although the name is not cited in text. Italic numbers give the page on which the complete reference is listed.

C

D

G

H

I

J

K

N

R

S

Z

Subject Index